Industrial

How to Plan, Install, and Maintain TCP/IP Ethernet Networks: *The Basic Reference Guide for Automation and Process Control Engineers*

Third Edition

Industrial Ethernet

How to Plan, Install, and Maintain TCP/IP Ethernet Networks: *The Basic Reference Guide for Automation and Process Control Engineers*

Third Edition

By Perry S. Marshall
and John S. Rinaldi

Notice

The information presented in this publication is for the general education of the reader. Because neither the author nor the publisher has any control over the use of the information by the reader, both the author and the publisher disclaim any and all liability of any kind arising out of such use. The reader is expected to exercise sound professional judgment in using any of the information presented in a particular application.

Additionally, neither the author nor the publisher has investigated or considered the effect of any patents on the ability of the reader to use any of the information in a particular application. The reader is responsible for reviewing any possible patents that may affect any particular use of the information presented.

Any references to commercial products in the work are cited as examples only. Neither the author nor the publisher endorses any referenced commercial product. Any trademarks or tradenames referenced belong to the respective owner of the mark or name. Neither the author nor the publisher makes any representation regarding the availability of any referenced commercial product at any time. The manufacturer's instructions on the use of any commercial product must be followed at all times, even if in conflict with the information in this publication.

Trademark Notices

Allen-Bradley, ControlLogix, FactoryTalk, PLC-5, Rockwell Automation, Rockwell Software, RSLinx, RSView, the Rockwell Software logo, and VersaView are registered trademarks of Rockwell Automation, Inc.

Other Trademarks

Microsoft, Windows, Windows ME, Windows NT, Windows 2000, Windows Server 2003, and Windows XP are either registered trademarks or trademarks of Microsoft Corporation in the United States and/or other countries.

Adobe, Acrobat, and Reader are either registered trademarks or trademarks of Adobe Systems Incorporated in the United States and/or other countries.

ControlNet is a registered trademark of ControlNet International.

DeviceNet is a trademark of the Open DeviceNet Vendor Association, Inc. (ODVA).

Ethernet is a registered trademark of Digital Equipment Corporation, Intel, and Xerox Corporation.

OLE for Process Control (OPC) is a registered trademark of the OPC Foundation.

All other trademarks are the property of their respective holders and are hereby acknowledged.

Printed in the United States of America.
10 9 8 7 6 5 4 3 2

ISBN: 978-1-945541-04-9

ISA
67 T. W. Alexander Drive
P.O. Box 12277
Research Triangle Park, NC 27709

Library of Congress Cataloging-in-Publication Data in process

This book is dedicated to the
Master Engineer,
whose works inspire and challenge
all designers. His creations are
beautiful, adaptable, robust,
and supremely equipped for their purpose.

Contents

About the Authors . . . . **xiii**

Acknowledgments . . . . **xv**

1.0 What Is Industrial Ethernet? **1**

1.1 Introduction 1
1.2 A Very Short History of Ethernet and TCP/IP 4

2.0 A Brief Tutorial on Digital Communication **7**

2.1 Digital Communication Terminology 9
Signal Transmission 9
Attenuation 9
Bandwidth 9
Noise 10
Message Encoding Mechanisms 10
Signal Encoding Mechanisms 11
Signaling Types 13
Error Detection 14
Checksum 14
Cyclic Redundancy Check 14

2.2 What's the Difference Between a Protocol and a Network? ... 15
Transmission/Reception of Messages ... 15
2.3 Basic Topologies ... 17
Hub/Spoke or Star Topology ... 18
Ring Topology ... 19
Mesh Topology ... 20
Trunk/Drop (Bus) Topology ... 20
Daisy Chain Topology ... 21
2.4 Arbitration Mechanisms ... 21
Contention ... 21
Token ... 22
Polling ... 22
2.5 LAN versus WAN versus VPN ... 22

3.0 Ethernet Hardware Basics ... 25

3.1 Ethernet Terminology ... 25
10BASE5: Thick Ethernet (Thicknet) ... 26
10BASE2: Thin Ethernet (THINNET) ... 26
10BASE-T: Twisted-Pair Ethernet ... 27
10BASE-F: Fiber-Optic Ethernet ... 29
Fast Ethernet ... 30
Gigabit Ethernet ... 35
3.2 Ethernet Hardware LEDs ... 37
3.3 Physical/Embedded Components: MAC, PHY, and Magnetics ... 37
3.4 Auto-Negotiation ... 39
3.5 Network Collisions and Arbitration: An Analogy ... 39
3.6 How the CSMA/CD Protocol Works ... 42
3.7 The Basic "Ethernet Design Rules" ... 45
3.8 "Would Somebody Please Explain This 7-Layer Networking Model?" ... 45
Layer 7: Application ... 46
Layer 6: Presentation ... 47
Layer 5: Session ... 47
Layer 4: Transport ... 47
Layer 3: Network ... 48
Layer 2: Data Link ... 48
Layer 1: Physical Layer ... 48

3.9 Connectors 49
IP67 Sealed Connector System for Industrial Ethernet 50
3.10 Pinouts 52
Ethernet DB-9 Connector 54
M12 "Micro" Connector for Industrial Ethernet 55

4.0 Ethernet Protocol and Addressing 57

4.1 A Little Bit of History 57
4.2 The Ethernet Packet and How Messages Flow on Ethernet 58
4.3 What Is the TCP/IP Protocol Suite? 61
4.4 TCP/IP Protocol Suite – IP Protocol 62
Why IP Addresses Are Necessary 65
The New Internet Protocol Version 6 66
Network ID versus Host ID 67
Legacy Address Classes 67
Today: Classless Subnet Masks 67
Assigning IP Addresses: Will Your Private LAN be Connected to the Internet? 69
Reducing the Number of Addresses Routers Must Advertise with "Supermasks" 70
4.5 TCP/IP Protocol Suite – TCP Protocol 71
4.6 TCP/IP Protocol Suite – UDP Protocol 74
4.7 Ports – How the TCP/IP Suite Is Shared Between Applications 75
4.8 Other TCP/IP Application Layer Protocols 76
DHCP 76
SNMP 77
TFTP 77
DNS 78
HTTP 78
FTP 78
Telnet 79
4.9 Popular TCP/IP Utilities 80
PING 80
Netstat 83

ARP . . . 83
The ARP Utility . . . 84

5.0 Basic Ethernet Building Blocks . . . 87

5.1 Devices . . . 88
Hubs . . . 88
Bridges . . . 89
Switches . . . 91
Routers . . . 92
Types of Routers . . . 92
Gateways . . . 94
Interface Cards . . . 94
5.2 Determinism, Repeatability, and Knowing if It's "Fast Enough" . . . 94
Achieving Determinism on Ethernet . . . 96
How Priority Messaging Works . . . 97
How Switches Determine Priority . . . 97
Drivers and Performance . . . 98

6.0 Network Health, Monitoring, and System Maintenance . . . 101

6.1 What Is It that Makes a Network Run Well? . . . 101
Monitoring . . . 102
Monitoring Switched Networks . . . 104
Documenting . . . 105
Troubleshooting . . . 106
6.2 Popular PC-Based Ethernet Utilities, Software, and Tools . . . 108

7.0 Installation, Troubleshooting, and Maintenance Tips . . . 111

7.1 Ethernet Grounding Rules . . . 111
Ethernet Grounding Rules for Coaxial Cable . . . 111
Twisted-Pair Cable Types . . . 112
Grounding for Shielded Twisted Pair . . . 113
Reducing Electromagnetic Interference (EMI) . . . 113
Switches Are Better than Hubs . . . 115
Better Cables Are Not Always Better . . . 115
Don't Skimp on Cables and Connectors . . . 116
Harsh Chemicals and Temperature Extremes . . . 116

7.2 When You Install Cable 116
7.3 How to Ensure Good Fiber-Optic Connections 118
Fiber-Optic Distance Limits 118
Full-Duplex Ethernet with Single-Mode Fiber 120

8.0 Ethernet Industrial Protocols, Fieldbuses, and Legacy Networks 121

8.1 The Two Most Important Points to Understand 123
8.2 Modbus and Modbus TCP 125
8.3 EtherNet/IP 130
8.4 PROFINET 137
8.5 FOUNDATION Fieldbus High-Speed Ethernet 140

9.0 Basic Precautions for Network Security 143

10.0 Power over Ethernet (PoE) 151

10.1 What Is PoE? 151
10.2 What Pins Are Used on the CAT5 Cable? 152
10.3 How Much Current Is Supplied? 153
10.4 What Are the Advantages to PoE? 154
10.5 How Do I Get Started with PoE? 155
10.6 Resources 156

11.0 Wireless Ethernet 157

11.1 A "Very" Short Technology Primer 157
11.2 Access Points 160
11.3 Mesh Networks 161
11.4 Security 161
11.5 The Advantages 163

12.0 Advanced Hardware Topics 165

12.1 Time Synchronization 165
12.2 Dual Ethernet Devices 168
12.3 Device Ring and Redundancy 170
12.4 Summary 173

13.0 The Internet of Things 175

13.1 Microsoft and the IoT 176

13.2 Amazon and the IoT 177
13.3 Oracle and the IoT 179
13.4 Summary 180

14.0 Factory Floor/Enterprise Communications 183

14.1 Tight versus Loosely-Coupled Systems 185
14.2 OPC UA for Factory-Enterprise Communications 188
14.3 Ten Things to Know about OPC UA 189
14.4 Reference 194
14.5 Summary 195

15.0 The Alphabet Soup of the Internet of Things 197

15.1 XML 197
What Is XML? 198
How Is XML Used? 199
Summary 199
15.2 MTCONNECT 200
Overview 201
Summary 202
15.3 HTTP 202
What Is HTTP? 203
How Is HTTP Used? 203
Summary 204
15.4 REST 204
What Is REST? 205
How Is REST Used? 207
Summary 208
15.5 MQTT 208
Overview 208
What Is MQTT? 208
What Are the Benefits of MQTT? 209
Summary 210
15.6 DDS 210
Overview 210
What Is DDS? 211
What Are the Benefits of DDS? 212
Summary 213

Index 215

About the Authors

Perry S. Marshall is an author, speaker, and consultant who advises technology companies on product definition, marketing, and new customer acquisition. He wrote *80/20 Sales & Marketing*, which applies the Pareto principle to marketing processes; *Ultimate Guide to Google AdWords*, the world's most popular book on internet advertising; and *Evolution 2.0: Breaking the Deadlock Between Darwin and Design*, which applies communication theory to biology. He has a BSEE from the University of Nebraska. You can contact him via his website, www.perrymarshall.com.

John S. Rinaldi has worked on industrial and building networking for more than 30 years as both a user and a product developer. His experience includes low-level device networking, sensor bus communications, industrial Ethernet, and SCADA systems. Current interests include remote monitoring, industrial device/enterprise communications, and machine-to-machine communications. His company, Real Time Automation, is a leader in industrial and building automation networking software, turnkey systems, and add-on networking hardware. He can be reached through the website at www.rtaautomation.com.

Acknowledgments

A special thanks to the following people who helped with the content of this book:

Deon Reynders of IDC (www.idc-online.com) is the author of *Practical TCP/IP & Ethernet Networking for Industry*, and he teaches IDC's excellent introductory Industrial Ethernet course for beginners. He proofread the manuscript.

Deepak Arora is a founder of Saltriver Infosystems (www.saltriver.com). He does technical support, market research, and sales of networked and wireless business systems. He proofread the manuscript and provided many illustrations.

Lynn August Linse (www.linse.org) is one of the automation industry's sharpest networking application engineers and programmers, who shares my vision for the convergence of networks and protocols. He helped by suggesting content improvements.

George Karones of Contemporary Controls (www.ccontrols.com) is a hardware developer and networking application specialist. He provided important information on embedded components.

Vivek Samyal (web@tannah.net) is an automation and web applications programmer and webmaster who specializes in database work and application-side programming. He supplied information on network analysis tools.

Laura, my wife, is considered by most to be a saint. She put up with me during the complex process of writing this book.

1.0 What Is Industrial Ethernet?

1.1 Introduction

Industrial Ethernet is the successful application of IEEE 802.3 standards with wiring, connectors, and hardware that meet the electrical noise, vibration, temperature, and durability requirements of factory equipment, as well as network protocols that provide interoperability and time-critical control of smart devices and machines.

Industrial Ethernet is a specialized, rigorous application of standard "office Ethernet" technology that adds any or all the following requirements:

- **Mission critical** – Downtime is much less tolerable in the factory than the office. When an office network goes down, you go get a cup of coffee and check your email later. When a factory goes down, you choke down your last mouthful of coffee, run into the plant, and fix the problem as fast as possible! The effects of downtime are less isolated in a manufacturing facility.
- **Harsh environment** – Factory equipment is not usually installed in air-conditioned hall closets. It's more likely

to be bolted to a robotic welder or oil rig. Temperature extremes and vibration threaten garden-variety hardware, cables, and connectors. Device selection, installation, and proper wiring practices are crucial.

- **Electrical noise** – Ordinary 110 VAC circuits are not the norm in factories. Industrial Ethernet devices are often used with high-current 480 VAC power lines, reactive loads, radios, motor drives, and high-voltage switchgear. Network communication must continue reliably despite these hazards.

- **Vibration** – Industrial Ethernet "smart devices" are, by definition, mounted on machines. Machines move and shake. Velcro and "telephone connectors" may not be up to the task.

- **Powered devices** – Certain devices must be powered by the network cable itself. Many automation devices operate at 24 VDC. Many devices are now powered directly from the Ethernet network.

- **Security** – The data in your factory is not necessarily more worthy of protection than the data in your office, but the threats are different. Factory equipment is vulnerable to hackers, of course, but accidental disruptions created by yourself or your staff are much more likely. Specific precautions must be taken.

- **Legacy devices** – Real automation systems are a mix of new, nearly new, old, older, and pre-Mesozoic Era equipment from incompatible vendors. Industrial Ethernet must link serial protocols, legacy networks, and fieldbuses.

- **Interoperability** – Ethernet devices must communicate with each other, with PCs, with enterprise business systems, and with cloud applications. The existence of an Ethernet jack is no guarantee of openness, interoperability, or compatibility. You must ask the right questions when making purchases.

- **Levels of priority** – Some machine-control information requires real-time, deterministic responses. Other data is much less urgent. It's important to recognize different priority levels for different kinds of data.
- **Performance** – Beyond physical robustness are subtle characteristics of software drivers, routers, and switches, such as hidden latency, jitter, limited connection, and behavior under erratic conditions.
- **Connectivity to other local area networks (LANs)** – Most Industrial Ethernet systems must be bridged to business intranets and the Internet. Serious problems can be introduced on both sides if this is not done with care.
- **The IT department versus the automation department** – Ethernet is precisely the place where two equally valid but conflicting views of "systems" and "data" come together. You must proceed with care to avoid a battle between company fiefdoms, all-out mutiny, or even a brand new pair of cement shoes.
- **Mastery of the basics** – No matter how good your equipment is, if you don't apply proper knowledge of Ethernet, Transmission Control Protocol/Internet Protocol (TCP/IP), and sound installation practices, your system will never work right.

Industrial Ethernet is a reference book that addresses each of these concerns and lays down the basic nuts and bolts of Ethernet and TCP/IP. After reading this book, you'll know the basics of the world's most popular network, you'll be able to plan Ethernet projects, and you'll know the right questions to ask when you talk to vendors.

Ethernet is the worldwide de facto standard for linking computers together. Ethernet connects hundreds of millions of computers and smart devices across buildings, campuses, cities, and countries. Cables and hardware are widely available and inexpensive ("dirt cheap" in the case of ordinary office-grade products), and software is written for almost every computing platform.

Ethernet is now a hot topic in automation, where industry-specific networks have dominated: PROFIBUS, DeviceNet, Modbus, Modbus Plus, Remote I/O, Genius I/O, Data Highway Plus, FOUNDATION Fieldbus, and numerous serial protocols over the electrical standards of EIA RS-232, RS-422, and RS-485.

In some cases, Ethernet is displacing these networks. In nearly all cases, Ethernet is being used in demanding installations alongside them. This book gives a basic understanding of Ethernet's strengths, weaknesses, fundamental design rules, and application guidelines. It addresses the unique demands of the factory environment, intelligent devices, and the most common automation applications and protocols. *Industrial Ethernet* provides basic installation and troubleshooting recommendations to help your projects work right the first time.

1.2 A Very Short History of Ethernet and TCP/IP

Ethernet originated at Xerox Palo Alto Research Center (PARC) in the mid-1970s. The basic philosophy was that any station could send a message at any time, and the recipient had to acknowledge successful receipt of the message. It was successful and in 1980 the DIX Consortium (Digital Equipment Corporation, Intel, and Xerox) was formed, issuing a specification, *Ethernet Blue Book 1*, followed by *Ethernet Blue Book 2*. This was offered to the Institute of Electrical and Electronics Engineers (IEEE, www.ieee.org), who in 1983 issued the *Carrier Sense Multiple Access with Collision Detection* (CSMA/CD) specification—their stamp of approval on the technology.

Ethernet has since evolved under IEEE to encompass a variety of standards for copper, fiber, and wireless transmission at multiple data rates.

Ethernet is an excellent transmission medium for data, but by itself falls short of offering a complete solution. A network protocol is also needed to make it truly useful, and TCP/IP has evolved alongside of Ethernet.

The big push toward TCP/IP came in the mid-1980s when 20 of the largest U.S. government departments, including the U.S. Department of Defense, decreed that all mainframes (read: expensive computers) to be purchased henceforth required a commercially listed and available implementation of UNIX to be offered. The department didn't necessarily need to use UNIX for the project at hand, but after "the project" was completed, the government wanted the ready option to convert this expensive computer into a general-purpose computer.

This soon meant that all serious computer systems in the world had relatively interoperable Ethernet and TCP/IP implementations. So IBM had Systems Network Architecture (SNA), TCP/IP, and Ethernet on *all* of its computers. Digital Equipment Corporation (DEC) had DECnet, TCP/IP, and Ethernet on all of its computers. Add a few more examples (Cray, Sun, CDC, Unisys, etc.) and you soon see that the only true standard available on all computers was a TCP/IP plus Ethernet combination.

Both from a historical view, as well as in today's industrial world, the TCP/IP plus Ethernet marriage is a key combination. Neither would have survived or prospered without the other.

2.0

A Brief Tutorial on Digital Communication

Digital communication is the transmission of data between two or more intelligent devices in a mutually agreed upon electronic format (e.g., binary, octal, EBCDIC, and ASCII). The following components are necessary to accomplish this:

- Data source
- Transmitting electronics
- Communications media
- Receiving electronics
- Data destination

The fundamentals of communication are the same, regardless of the technology. Confusion about any aspect can usually be helped with direct analogies to more familiar modes of communication such as multiple people engaged in a conversation around the dinner table, telephones, CB radios, or Morse code.

Communication standards define agreement on key details:

1. **Physical Connections:** How the signal gets from one point to another
    - The actual form of the physical connections
    - Signal amplitudes, grounding, physical media (coaxial cable, fiber, twisted pair, etc.)
    - Transmitting, receiving, and isolation circuitry
    - Safe handling of fault conditions such as miswiring or shorts to ground
2. **Protocol:** How messages are formatted and delivered
    - Byte ordering
    - Header and trailer bytes
    - Flow control
    - Error detection
    - Prioritization
    - Retries and missing message detection
    - Synchronization: coordinating the timing of message events
3. **Encoding:** How the series of 1s and 0s that form the message are organized. Message encoding describes how the receiver will assemble the message packet of the protocol from the 1s and 0s. Electrical encoding describes how the 1s and 0s are organized on the wire to overcome various electrical constraints (more on this later).
    - Message encoding standards (includes formats like ASCII, XML, or simply binary)

- Signal encoding on the wire (includes encoding standards like Manchester, RZ, and NRZ encoding)

2.1 Digital Communication Terminology

Signal Transmission

In the literal sense, all communication signals in a transmission are analog. Whether it's a digital pulse train on a wire or laser light on a fiber-optic line, the physical nature of the media impose *attenuation* and *bandwidth* limits on the signal.

Attenuation

Loss of signal amplitude for any reason is called *attenuation*. When someone shouts to you from down the street, the farther away they are, the more attenuated their voice becomes—meaning loudness and clarity are lost. Attenuation is, among other things, a function of frequency and distance.

In combination with noise, attenuation dictates your ability to move data over long distances. The various digital communications standards are impacted by attenuation to different degrees. It is part of the reason that RS-232 is limited to 15 m and RS-485 can go up to 1000 m. Whether you're talking about sounds in the air or signals on a wire, attenuation is normally expressed in decibels.

Bandwidth

The *bandwidth* of a medium is its ability to move useful data over time—for example, 400,000 words per second or 20 messages per second. The transmission speed depends on the bandwidth, while noise, data retries, and other "overhead" eat into the bandwidth.

Transmission speed is limited by the ability of the medium to rapidly change states between 1 and 0. When a transmitter changes from 0 to 1, there is some delay in the remote receiver noticing this change. Noise filtering causes the remote receiver to ignore the leading edge of the change, and capacitance and other electrical properties resist the change within the media.

Noise

Noise is any unwanted signal that interferes with data transmission. Noise on a network can be created by external sources such as power lines, radios, welders, switchgear, cellular telephones, etc. and is induced by coupling of magnetic and/or electric fields. The most typical measurement of noise is the signal-to-noise ratio, expressed in decibels. The oft-cited advantage of digital communication is this: if noise levels are kept below a certain threshold, it does not affect communication. In reality, noise is usually sporadic, sometimes affecting messages and sometimes not. Tips for reducing noise in Ethernet networks are given in Chapter 7, "Installation, Troubleshooting, and Maintenance Tips."

Message Encoding Mechanisms

Message encodings mechanisms define standard ways of representing messages to a receiver. To send a message to another node, it must be represented in a way that the target system can understand it. The message encoding defines how the message is represented for consumption by the receiver.

Selecting a data encoding is an important application consideration. The data encoding affects the system performance and how easy it might be to pass data to other systems. Data encoded using a standard encoding provides for much easier integration with standard systems. Data encoded in binary using a protocol specific encoding generally offers higher performance but sometimes requires a more custom integration.

Most technologies use a single binary encoding that is fixed and unchanging. Many communication protocols simply encode the message bytes as binary data and transmit those bytes, which is the traditional way to do message encoding. In that kind of message encoding, a 16 bit integer (2 bytes) might be encoded as high byte followed by low byte. Both sender and receiver would have to know how each type of message field is encoded for successful message communication.

Another kind of encoding you might recognize is Extensible Markup Language (XML) encoding. XML encoding is a standard way of encoding messages in IT systems in which values are encoded in ASCII and surrounded by a tag name in angled brackets like this:

<function code>15</function code>

<Start Address>100</Start address>

Many computer systems are able to receive and interpret a message encoded as XML. An encoding like XML provides ease of integration with IT applications while a proprietary binary encoding is much less verbose and more efficient for smaller-resourced embedded devices.

Signal Encoding Mechanisms

When you think of 1s and 0s on a wire, it's intuitive to assume that the data appears on the wire exactly as it does in the packet. Actually this is seldom the case. During a long string of continuous 0s or 1s (which is certain to happen from time to time), the receiver may think the connection has been lost and can lose synchronization with the transmitter. There are a variety of specific mechanisms for preventing problems like this, with tradeoffs between noise immunity, bandwidth, and complexity. The following are the most common formats:

- **Manchester (used in 10 Mb Ethernet)** – The state of a bit is represented by a *transition* between V+ and V- in the middle of the bit. Here, 1s are represented by a downward swing from V+ to V-; 0s are represented by an upward swing from V-to V+. There is ALWAYS a transition, regardless of the actual bit sequence. Advantage: The receiver and transmitter clocks are always synchronized. Disadvantage: This scheme uses twice as many transitions as bits.

- **RZ (Return-to-Zero)** – The signal state is determined by the voltage during the first half of each bit, and the signal returns to a resting state ("zero") during the second half of each bit.

- **NRZ (Non-Return-to-Zero)** – This is simply a direct, intuitive, "1 = high, 0 = low" designation with no further coding.

- **MLT-3** – A three-level algorithm (i.e., high, zero, and low voltages) that changes levels only when a 1 occurs. Not self-clocking.

- **Differential Manchester** – Bit value is determined by the presence or absence of a transition at the beginning of a bit interval; clocking is provided via a mid-interval transition.

- **4B/5B (4 bit/5 bit)** – Every 4 bits is represented as a 5-bit code that never has more than three 0s in a row. This prevents long sequences of 0s or 1s with only a 25% penalty in bandwidth, in contrast to the 100% penalty of Manchester.

Signaling Types

The rubber meets the road in an Industrial Ethernet network when the data from a device is transferred to the communication media. The process of transferring data to the wire is called *signaling* and there are two basic types: baseband and broadband.

Baseband signaling is digital. The 1s and 0s of a message are transmitted over the media as a sequence of voltage pulses. If you remember the old time Westerns, the clerk in the telegraph office would tap out a telegram using digital signaling. The major limitation to digital signaling is that only a single message can be transmitted at a time. If there are two clerks in that telegraph office, the second clerk must wait for the first clerk to finish before sending the next message. A second limitation is that the digital voltages are easily attenuated as the distance increases. Communication over long distances requires repeaters and is almost impractical due to the number of repeaters required.

Broadband signaling doesn't have this restriction. Broadband signaling uses analog carrier to transport data. Multiple carriers each containing data is analogous to a cable-TV system. The cable carries many television programs all on the same wire and all at the same time. You select a different carrier and a different data stream (a television program) by switching from Channel 9 to Channel 10.

Table 2-1. Baseband versus Broadband Signaling Characteristics

Baseband	Broadband
Digital signaling	Analog signaling
Limited distance	Long distances
Bus-oriented applications such as RS-232 and controller area networking (CAN)	Used for both bus and tree topologies such as token ring and ethernet
Bi-directional	Uni-directional
Single message-oriented	Multiple carrier signals with multiple independent data streams
Often uses Manchester encoding	No encoding of digital signals

Error Detection

The simplest mode of error detection is "echoing back" the message just sent. However this consumes double bandwidth. Plus if there's an error, it's impossible to tell whether it was the original or the copy that was corrupted.

Checksum

The *checksum* calculation is effective for small amounts of data. An algorithm converts the data to bits that are appended to the data and transmitted. The receiver does the same calculation on the same data; if its own result does not match the original checksum, a retransmit request is submitted. For a single byte of data, a 1-bit checksum (parity bit) is sufficient.

Cyclic Redundancy Check

Long messages require a more sophisticated, more accurate detection method. *Cyclic Redundancy Check* (CRC) views the entire message block as a binary number, which it divides by a special polynomial. The result is a remainder, appended to the message just like a checksum. CRC calculation is performed in real time by logic gates at the hardware level.

Not only are the above mechanisms employed in hardware, they are also employed in higher-level protocols. TCP/IP employs its own error-detection mechanisms to further guarantee successful message delivery.

2.2 What's the Difference Between a Protocol and a Network?

The distinction between the physical network itself and the protocol that runs on that network is sometimes blurred. It's important to clarify: The network itself consists of the physical components and message-transmission hardware. Protocols are binary "languages" that run on the networks.

Strictly speaking, the terms *Ethernet, RS-232, RS-422, and RS-485,* for example, refer to the network physical wiring and message-transmission components only (see layers 1 and 2 of the ISO/OSI model in Chapter 3).

Many different protocols are used on Ethernet. TCP/IP, File Transfer Protocol (FTP), Hypertext Transfer Protocol (HTTP), NetBEUI, AppleTalk, and Modbus are protocols only; they can run on many different physical networks.

Transmission/Reception of Messages

Simplex. Simplex is one-way communication via a single channel. A radio or TV tower is a simplex transmitter; a radio or TV is a simplex receiver.

Duplex. Duplex is two-way communication.

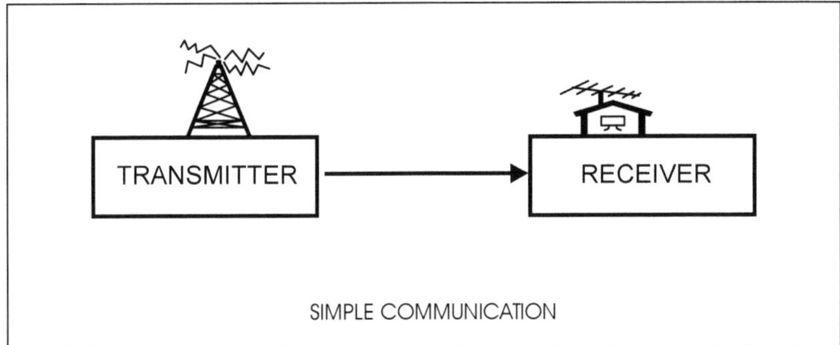

Figure 2-1. Simplex Communication

Half Duplex. Half-duplex communication is when both stations (e.g., Walkie-Talkie or CB radio) can transmit and receive but they cannot do it simultaneously.

In half-duplex communication, only one party can have control of the channel at any one time. This necessitates an arbitration mechanism to determine who has control of the channel. This is called *contention*.

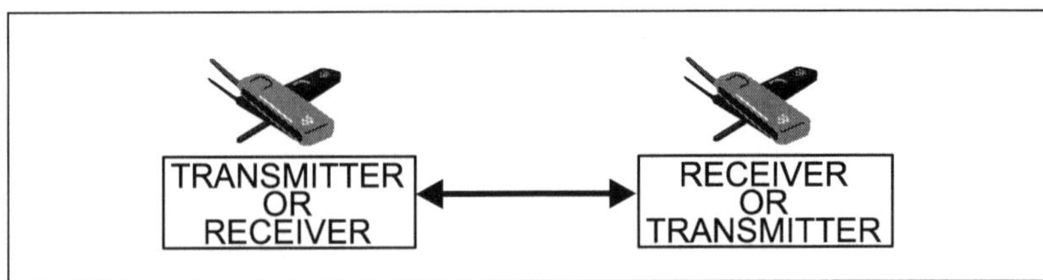

Figure 2-2. Half-Duplex Communication

Full Duplex. Full duplex is two-way communication with two communications channels so that both stations can receive and transmit simultaneously. A telephone is full duplex because it allows both parties to talk and listen at the same time.

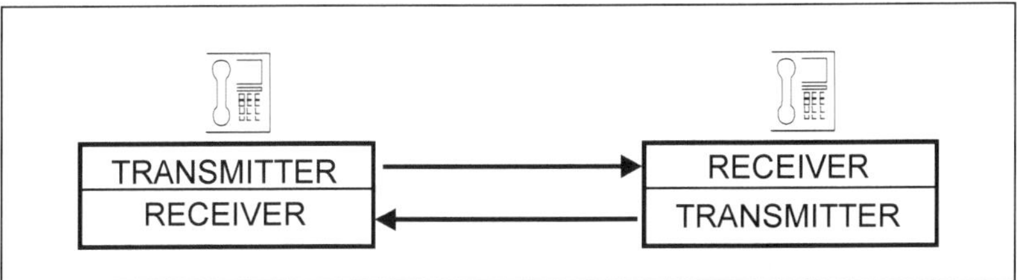

Figure 2-3. Telephone Conversation is Full Duplex (at least to the extent that a person can talk and listen at the same time)

In Ethernet, half-duplex communication requires the use of CSMA/CD arbitration; full duplex eliminates collisions altogether but requires separate transmit and receive paths between each device. Ethernet always has separate Tx/Rx paths. Full duplex in Ethernet requires not only separate paths, but only two nodes in a collision domain.

2.3 Basic Topologies

Topology is an important choice in system design. It dictates what kind of physical arrangement of devices is possible. Figures 2-5 through 2-9 show what topologies are supported by each flavor of Ethernet.

A network is an electrical transmission line. At high speeds, each bit is short compared with the network length. If you could physically see the packets traveling across the wire, each bit would have a length, similar to the wavelength of sound or light. Data propagates on wire at about 2/3 the speed of light.

When a wave reaches the end of a medium, it is reflected, transmitted, and/or absorbed. The shorter the bits in relation to the network, the more likely that reflections will cause errors.

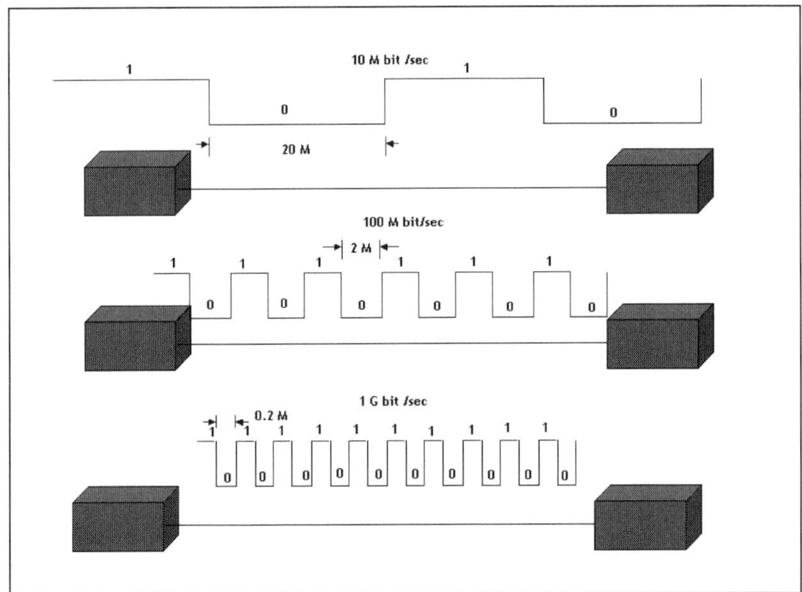

Figure 2-4. As Bit Rate Increases, the Physical Length of Each Bit Decreases

Network speed	10 Mbps	100 Mbps	1 Gbps
Distance signal travels in the duration of 1-bit time	20 m	2 m	0.2 m

For high-speed networks, the simplest way to minimize reflections is to have only one node at each end of a wire, with proper impedance termination (wave absorption) at each end. If each node has terminating resistors matching the cable impedance, reflections are minimized.

Hub/Spoke or Star Topology

A *hub/spoke* or *star topology*, where every segment has dedicated transmitters and receivers, offers high performance because reflections and impedance mismatches are minimal. This is the topology used by all the Ethernet formats except 10BASE2 and 10BASE5.

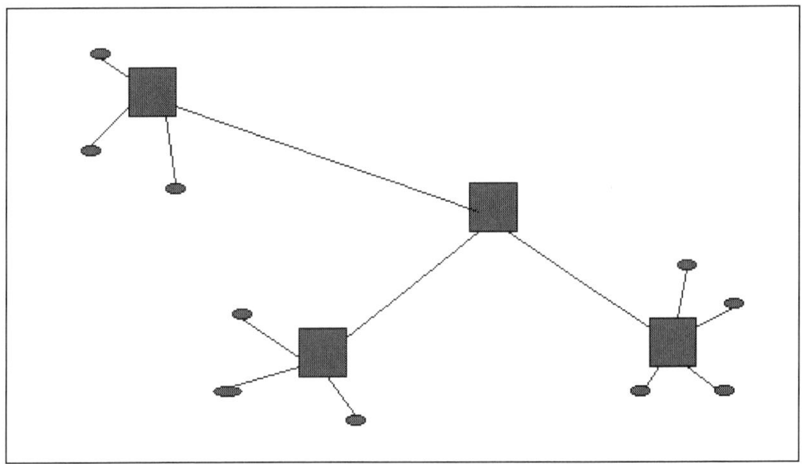

Figure 2-5. Star Topology

Ring Topology

Ring topology could be looked at as a variation on hub/spoke. It is similar in the sense that each segment has dedicated transmitters and receivers. However, the data itself is passed around in a circle, and it is stored and forwarded by each node—an important distinction.

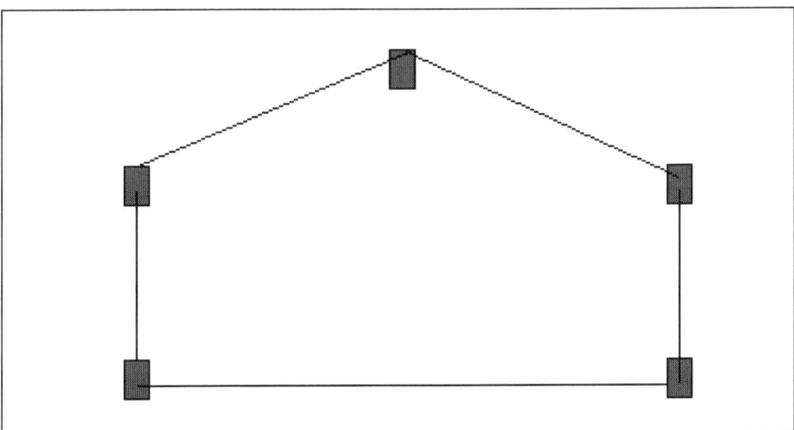

Figure 2-6. Ring Topology

Mesh Topology

Mesh topology is point-to-point like star and ring, but has a minimum of two paths to and from each network node. This provides redundancy but introduces significant cost and installation effort.

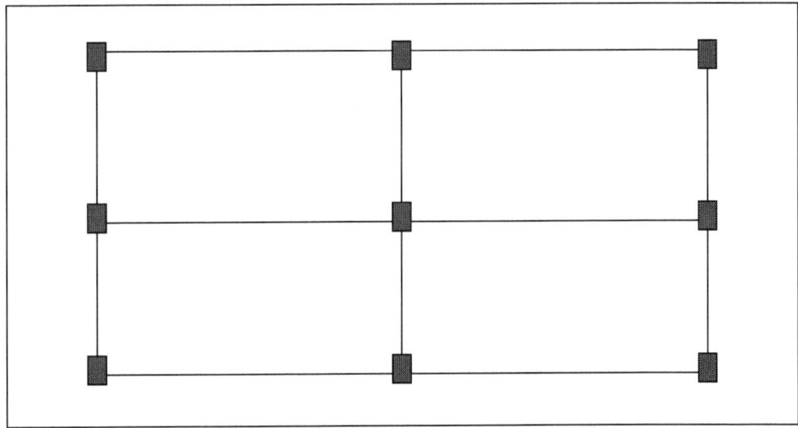

Figure 2-7. Mesh Topology

Trunk/Drop (Bus) Topology

Trunk/drop topology, also known as *bus* or *multidrop topology,* puts multiple nodes along the distance of the cable, with spurs or "Tees" inserted wherever a node is needed. Each spur introduces some reflections, and there are rules governing the maximum length of any spur and the total length of all spurs. 10BASE5 is a trunk/drop implementation of Ethernet.

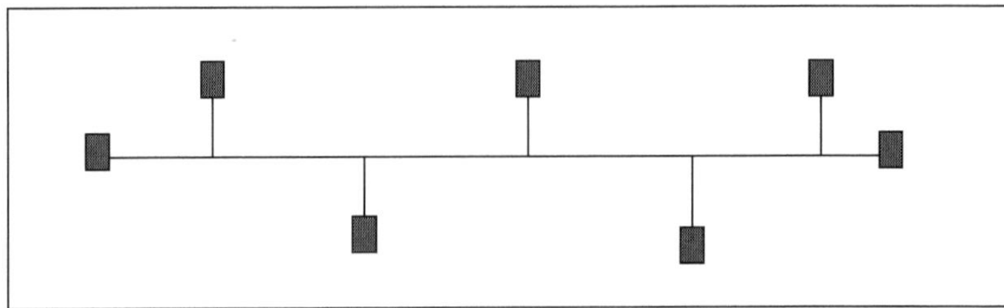

Figure 2-8. Bus Topology

Daisy Chain Topology

A variation on trunk/drop is the *daisy chain*, where spur length is reduced to zero. High bandwidth signals have fewer problems in a daisy chain than trunk/drop because of fewer reflections. RS-485 is an example of a daisy chain; controller area networks (CANs) like DeviceNet use trunk/drop. 10BASE2 is a daisy chain implementation of Ethernet; the drop length is effectively zero.

Star topologies have a nice advantage over trunk/drop: Errors are easier to assign to a single segment or device. The disadvantage is that some physical layouts (e.g., long conveyor systems with evenly spaced nodes) are difficult to implement on a star; trunk/drop or daisy chain are better for this layout.

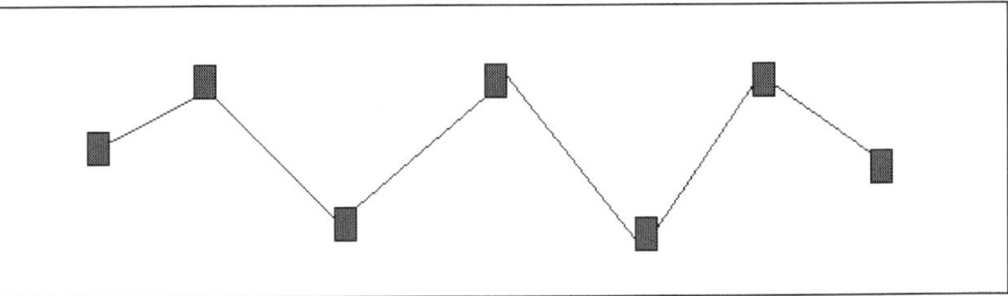

Figure 2-9. Daisy Chain Topology

2.4 Arbitration Mechanisms

There are three basic methods of arbitrating between competing message sources:

Contention

Contention is similar to a group of people having a conversation where all are listening, one can speak at any given time, and when there is silence another can speak up. Two or more may interrupt the silence and then all but one must back off and wait their turn.

Token

Token messaging is when each device receives a sort of token or "turn to speak" and can transmit only while it is in possession of that token. The token is then passed to someone else who now can transmit. Since Ethernet is not token-based, not much space will be given to this topic. There are many possible rules for passing the token, but often it is passed in a cyclic fashion from one device to the next.

Polling

Polling is when one device is "in charge" and asks each device to surrender its data in turn. Polling systems are often deterministic but do not allow urgent messages to be prioritized over other messages.

2.5 LAN versus WAN versus VPN

Local area networks (LANs) transmit data at high speed over a limited area. A single Ethernet system on 10BASE-T or 100BASE-T is a typical LAN architecture. Such a system is limited in geography by the maximum number of hubs/switches (see Chapter 3), and propagation delays are in the 1-ms range and below.

Wide area networks (WANs) link LANs together over large distances. WANs usually use publicly available communication links from telecommunication providers. These links might consist of combinations of fiber, telephone, radio, and satellite links. Within a WAN, gateways often buffer packets until messages are complete, then forward them to the receiving computer. This causes propagation delays, and WANs are often unsuitable for real-time applications.

Virtual private networks (VPNs) link LANs via the Internet. Since data is then visible to others, data encryption is used to keep messages private. VPNs are extremely popular in companies with facilities in multiple locations and in companies that have remote or traveling employees. VPNs can extend through dial-up modem connections with the appropriate software installed on the dialing PC.

3.0 Ethernet Hardware Basics

3.1 Ethernet Terminology

The many formats of Ethernet cabling are described with rather unfriendly shorthand terminology. IEEE's Ethernet naming convention works like this:

- The first number (10, 100, 1000) indicates the transmission speed in megabits per second.
- The second term indicates transmission type: BASE = baseband; BROAD = broadband.
- The last number indicates segment length. A 5 means a 500 m segment length from original Thicknet.

Tip 1 – **You might assume that the 2 in 10BASE2 indicates a 200 m segment length, but don't be too literal.** Actually 10BASE2 supports 185 m, or 300 m running point-to-point without repeaters.

- In the newer standards, IEEE used letters rather than numbers. The T in 10BASE-T means Unshielded Twisted-Pair cables. The T4 in 100BASE-T4 indicates four pairs of Unshielded Twisted-Pair cables.

10BASE5: Thick Ethernet (Thicknet)

10BASE5 is the original IEEE 802.3 Ethernet. 10BASE5 uses thick yellow coaxial cable with a 10-mm diameter. The cable is terminated with a 50-ohm 1-W resistor. One hundred stations maximum per segment are allowed.

10BASE5 uses trunk/drop topology. Stations are connected with a single coaxial cable. The maximum length of one segment is 500 m, limited by the quality of the cable itself.

A network interface card (NIC) is attached with a 15-pin D-shell connector to a short attachment unit interface (AUI) cable, which in turn connects to a media attachment unit (MAU) and links to the coaxial cable by means of a *vampire connector* that pierces the cable. The MAU contains the actual transceiver that connects to the coaxial cable.

For proper CSMA/CD operation, the network diameter for 10BASE5 is limited to 2500 m, consisting of five 500-m segments with four repeaters.

10BASE2: Thin Ethernet (THINNET)

10BASE2 resembles 10BASE5. It was introduced to reduce the cost and complexity of installation. It uses RG-58 50-ohm coaxial cable that is cheaper and thinner than that used for 10BASE5, hence the name Cheapernet or Thinnet, which is short for "thin Ethernet."

10BASE2 integrates the MAU and the transceiver/AUI cable onto the NIC itself, with a Bayonet Nut Connector (BNC) replacing the AUI or D-15 connector on the NIC.

Lower cable quality means reduced distance. The maximum length of a 10BASE2 segment is 185 m. 10BASE2 supports 30 nodes per segment and keeps the four repeater/five segment rule. So the maximum network diameter of 5 segments × 185 m = 925 m.

Thinnet became popular and replaced thick Ethernet as an office cabling solution.

10BASE-T: Twisted-Pair Ethernet

In 1990, IEEE approved 802.3i 10BASE-T, a new media that is completely different from coax. 10BASE-T uses two pairs of unshielded twisted-pair (UTP) telephone-type cable: one pair of wires to transmit data, and a second pair to receive data. It uses eight conductor RJ-45 connectors.

The topology is star instead of trunk/drop, with only two nodes per segment allowed: Station to repeater, repeater to repeater, or station to station with a *crossover cable*, which is needed to cross the transmit and receive lines.

The maximum length of a segment is 100 m, which follows the EIA/TIA-568-B wiring standard. Repeater-repeater links are also limited to a maximum of 100 m.

10BASE-T uses the four repeater/five segment rule from 10BASE5 and 10BASE2. So a 10BASE-T LAN can have a maximum diameter of 500 m.

Like 10BASE2 and 10BASE5, 10BASE-T uses Manchester encoding. IT uses +V and –V voltages with differential drivers. The signal frequency is 20 MHz, and Category 3 or better UTP cable is required.

10BASE-T has a link integrity feature that makes installing and troubleshooting much easier. Devices on each end of the wire transmit a "heartbeat" pulse. Both the hub and the NIC look for this signal when connected. The presence of a heartbeat means a reliable connection is in place.

Most 10BASE-T devices have a light-emitting diode (LED) that indicates whether the link is good.

Tip 2 **– You should start troubleshooting wiring problems by looking at the state of the link LED at both ends of the wire.**

Most 10BASE-T equipment combines the functions of the MAU in the NIC or the hub itself.

In terms of bandwidth, coaxial cable is superior to UTP cable. However, UTP cabling and star topology are a real advantage because (1) in a bus topology, a problem at one node can take down the whole network, whereas a star topology makes it easier to isolate problems; (2) with the low cost of hubs and switches, star topology is still cost-effective; and (3) the existing cabling used in telecommunications equipment, especially CAT3 cable, could be used.

The star-shaped, planned, and structured wiring topology of telecommunications with 10BASE-T is different from and superior to the single-point-of-failure method of 10BASE5 and 10BASE2.

10BASE-F: Fiber-Optic Ethernet

10BASE-F actually refers to three different sets of fiber-optic specifications:

- 10BASE-FL (FL means "Fiber Link") replaces the older Fiber Optic Inter-Repeater Link (FOIRL) spec and is backward-compatible with existing FOIRL devices. It is the most popular 10 Mbps fiber standard and connects data terminal equipment (DTE), repeaters, and switches. Equipment is available from many vendors.
- 10BASE-FP and 10BASE-FB are dead. "P" stands for passive and "B" stands for backbone.

10BASE-F comes from the FOIRL specification of 1987, which linked repeaters using an extended distance fiber-optic link.

10BASE-F has twin strands of single-mode or multimode glass fiber, using one strand to transmit and the other to receive. Multimode fiber (MMF) of 62.5-/125-micron diameter is most often used with 10BASE-F to carry infrared light from LEDs. The specified connectors are IEC BFOC/2.5 miniature bayonet connectors, best known as *ST connectors*. SC and ST connectors are extremely popular in 10BASE-F.

Segment length for 10BASE-F ranges from 400 to 2000 m with a maximum of five segments on one collision domain.

10BROAD36 uses radio frequency transmission to carry data. This permits multiple channels to operate simultaneously on the same cable. 10BROAD36 is essentially dead, and no 100 Mbps version exists.

Fast Ethernet

100BASE-T. 100BASE-T is basically 10BASE-T with the original Ethernet media access controller (MAC), at 10 times the speed. The 100BASE-T allows several physical layer implementations. Three different 100BASE-T physical layers are part of IEEE 802.3u: two for UTP and one for multimode fiber. Just like 10BASE-T and 10BASE-F, 100BASE-T requires a star topology with a central hub or switch.

IEEE 802.3u contains three new physical layers for 100 Mbps Ethernet:

- **100BASE-TX** – Two pairs of Category 5 UTP or Type 1 shielded twisted-pair (STP) cabling; this is the most popular for *horizontal* connections. Uses two strands of 62.5-/125-micron Fiber Distributed Data Interface (FDDI) cabling.
- **100BASE-FX** – Two strands of multimode fiber; this is the most popular for *vertical* or backbone connections.
- **100BASE-T4** – Four pairs of Category 3 or better cabling; not common. 100BASE-T4 was part of IEEE 802.3u and was intended to capitalize on the huge installed base of Category 3 voice-grade wiring. It was a flop because T4 products only started shipping a year after the standard was approved; TX products got a head start.

See Table 3-1 for distance capabilities.

Table 3-1. Ethernet Physical Layer Characteristics

Format	Data rate	Max segment length *	Max nodes per segment	Topology	Media	Connectors	Encoding	Notes
10BASE-T	10 Mbps half duplex 20 Mbps full duplex	100 m Max network length = 100 m node to hub	2	Star	Category 3, 4, or 5 UTP cable with two pairs of voice-grade/telephone twisted pair, 100 ohms	8-pin RJ-45 style modular jack; industrial variants include M18, M12, and DB-9	Manchester	Most popular 10 m format
10BASE2 "Thinnet" or "Cheapernet"	10 Mbps half duplex only	185 m Max network length = 925 m = 5 × 185 m	30	Bus with drops. Minimum spacing between nodes = 0.5 m, max drop length = 4 cm	5 mm "thin" coax, e.g., RG58A/U or RG58C/U, Belden 9907 (PVC), and 89907 (plenum); 50 ohms	BNC "T" coax connectors, barrel connectors, and terminators	Manchester	5 cm min bend radius; may not be used as link between 10BASE5 systems
10BASE5 "Thicknet"	10 Mbps half duplex only	500 m (50 m max AUI length) Max network length = 2800 m = 5 × 500 m segments + 4 repeater cables + 2 AUI cables	100	Bus with drops	10 mm ("thick") coax, e.g., Belden 9880 (PVC) and 89880 (plenum); bend radius min 25 cm; 50 ohm media and termination	N-type coaxial connectors, barrel-style insulation displacement connectors and terminators	Manchester	MAU links trunk to NIC via AUI cable; taps must be spaced at 2.5 m intervals; ground at one end of cable**
10BROAD36	10 Mbps half duplex only	1800 m single segment; 3600 m total for multiple segments			75-ohm CATV broadband cable		Modulated RF	Dead

Table 3-1. Ethernet Physical Layer Characteristics (continued)

Format	Data rate	Max segment length *	Max nodes per segment	Topology	Media	Connectors	Encoding	Notes
10BASE-FL	10 Mbps half duplex; 20 Mbps full duplex	2,000 m	2	Star	2 MMF cables, RX and TX, typically 62.5/125 fiber, 850 nm wavelength	BFOC/2.5, also called "ST"	Manchester	Uncommon
100BASE-TX	100 Mbps half duplex; 200 Mbps full duplex	100 m	2	Star	2 pairs of Category 5 UTP cabling; 100-ohm impedance (optionally supports 150-ohm STP)	RJ-45 style modular jack (8 pins) for UTP cabling (optionally supports 9-pin D-shell connector for STP cabling)	4B/5B	Most popular 100 m format IEEE 802.3u
100BASE-FX	100 Mbps half duplex; 200 Mbps full duplex	Half duplex: 412 m; full duplex: 2,000 m	2	Star	2 MMF optical channels, one for TX, one for RX. Typ. 62.5/125 MMF, 1300-nm wavelength	Duplex SC, ST, or FDDI MIC connectors	4B/5B	IEEE 802.3u
100BASE-T4	100 Mbps half duplex only	100 m	2	Star	Category 3, 4, or 5 UTP (uses 4 pairs or wires); 100 ohm	RJ-45 style modular jack (8 pins)	8B/6T	Uncommon IEEE 802.3u Useful where existing CAT3 telecom cables are available

Table 3-1. Ethernet Physical Layer Characteristics (continued)

Format	Data rate	Max segment length *	Max nodes per segment	Topology	Media	Connectors	Encoding	Notes
1000BASE-LX	1,000 Mbps half duplex; 2,000 Mbps full duplex	Half-duplex MMF and SMF: 316 m; full-duplex MMF: 550 m; full-duplex SMF: 5000 m; 10 micron SMF: 3000 m max segment length	2	Star	2 62.5/125 or 50/125 multi-mode optical fibers (MMF), or 2 10-micron single-mode optical fibers (SMF), 1,270 nm to 1,355 nm light wavelength	duplex SC connector	8B/10B	803.z Multimode: longer-building backbones; Single mode: campus-wide backbones
1000BASE-SX	1,000 Mbps half duplex; 2,000 Mbps full duplex	Half-duplex 62.5/125: 275 m; half-duplex 50/125: 316 m; full-duplex 62.5/125: 275 m; full-duplex 50/125: 550 m	2	Star	2 62.5/125 or 50/125 MMF, 770 to 860 nm	duplex SC connector	8B/10B	
1000BASE SX	1,000 Mbps half duplex; 2,000 Mbps full duplex	Half duplex: 25 m; full duplex: 25 m 62.5/125 MMF full duplex: 260 m	2	Star	Specialty shielded balanced copper jumper cable (“twinax” or “short haul copper”)	9-pin shielded D-subminiature connector, or 8-pin ANSI Fiber Channel Type 2 (HSSC) connector	8B/10B	802.3z Intended for short backbones

Table 3-1. Ethernet Physical Layer Characteristics (continued)

Format	Data rate	Max segment length *	Max nodes per segment	Topology	Media	Connectors	Encoding	Notes
1000BASE-T	1,000 Mbps half duplex; 2,000 Mbps full duplex	100 m (328 ft)	2	Star	4 pairs of CAT5 or better cabling, 100 ohms	8-pin RJ-45 connector	PAM5	802.3ab Replace existing 10/100BASE-T runs in floors of buildings
1000BASE-CX		25 m	2	Star	STP copper Twinax, 150 ohms	DB-9 or HSSC	8B/10B	802.3z Short jumper connection in computer rooms or switching closets. Common ground required on devices at both ends of the cable

AUI = attachment unit interface; MAU= Medium attachment unit; MMF = Multimode fiber; NIC = Network interface card; SMF = Single-mode fiber; STP = Shielded twisted pair; UTP = Unshielded twisted pair; 62.5/125 means 62.5-micron fiber core with 125-micron outer cladding.

* For best results, keep segment length at least 20% shorter than recommended maximum.

** Ground should be made at one and only one point in a single link.

Maximum transmission path rules: 5 segments, 4 repeaters, 3 coax segments, 2 link segments OR 5 segments, 4 repeaters, 3 link segments, 2 coax segments.

Gigabit Ethernet

1,000 Mb Ethernet is just Fast Ethernet on steroids. 100BASE-T was wildly successful and it was only a matter of time before the data rate would be increased again. There are differences in the physical layers, network design, and minimum frame size. IEEE 802.3z, approved in 1998, includes the Gigabit Ethernet MAC, and three physical layers. Gigabit uses 8B/10B encoding. Gigabit encompasses three physical standards:

- 1000BASE-SX Fiber
- 1000BASE-LX Fiber
- 1000BASE-CX Copper
- 1000BASE-T

Some engineers wanted Gigabit Ethernet to be full duplex only, but CSMA/CD was kept. One of the reasons for keeping it was that it reduced the amount of redesign in migrating to 1000 Mb chips.

To make CSMA/CD work at 1 GHz, the slot time was increased to 512 bytes, as opposed to 64 bytes for 10 and 100 Mbps Ethernet. This is the allowable time in which the transmitter "holds the floor" and the complete frame is transmitted. If the transmitted frame is smaller than 512 bytes, a carrier extension is added to the end of the frame. The carrier extension resembles the PAD (see later in this chapter) that is added to the end of the data field inside the frame. Carrier extension, however, adds after the CRC and does not actually form part of the frame.

If most of the messages on a gigabit network are short, overhead makes the network extremely inefficient. So Gigabit Ethernet includes a feature called *burst mode.* A station can continuously transmit multiple smaller frames, up to a maximum of 8,192 bytes. This is done such that the transmitting node has continuous control of the media during the burst.

- **1000BASE-SX: Horizontal Fiber** – 1000BASE-SX is for low-cost, short-backbone, or horizontal connections (S stands for "short"). It has the same physical layer as LX and uses inexpensive diodes and multimode fiber. Distance ranges from 220 to 550 m, depending on the type of fiber.
- **1000BASE-LX: Vertical or Campus Backbones** – 1000BASE-LX is for longer-backbone and vertical connections (L stands for "long"). LX can use single-mode or multimode fiber. It requires more expensive optics. Segment length is 5000 m with single-mode fiber. For full-duplex multimode, the distance is 550 m. IEEE specifies the SC style connector for both SX and LX.
- **1000BASE-CX: Copper-Twinaxial Cabling** – 1000BASE-CX (C stands for *copper* or *cross-connect*) links hubs, switches, and routers in closets. Copper is preferred because it is faster to wire a connection with copper than with fiber. 150-W twinaxial cable is specified. Maximum length is 25 m for half or full duplex. Two connectors are used with 1000BASE-CX: The High-Speed Serial Data Connector (HSSDC) and the 9-pin D-subminiature connector, used for token ring and the 100BASE-TX STP.

***Caution:** You should be aware that Fast Ethernet and Gigabit Ethernet systems on copper media are extra susceptible to electrical noise for two reasons: (1) voltage levels are lower and thus more easily corrupted by noise, and (2) bit times are extremely short; a noise spike does not have to last very long to corrupt an entire frame.*

***Tip 3* – Take extra care to select high-quality cables and avoid routing them through electrically noisy areas if possible.**

3.2 Ethernet Hardware LEDs

Most Ethernet NIC cards, hubs switches, and other hardware have two LEDs:

1. The Link LED indicates that a reliable physical connection is established between the device and another device. This LED is the first thing you should check when something does not appear to be working correctly.
2. The TX LED turns on when the device is transmitting data. Some devices also have an RX LED that indicates data is being received.

***Tip 4* – These LEDs are the first thing you check when there appears to be a problem.**

3.3 Physical/Embedded Components: MAC, PHY, and Magnetics

At the lowest level, an Ethernet interface typically is made of four components. The first two are usually combined:

- The first component is the *media access controller* (MAC), for example, the popular AMD 79C960 chip and its derivatives. The controller assembles and disassembles Ethernet frames and provides an interface to external data.

- Internal to most MACs (separate in some cases) is an *intermediate interface*. It allows independence from the different types of transmission media (copper, fiber). For 10 Mbps Ethernet, the interface is called an *attachment unit interface* (AUI). In 10BASE5 systems, this is an external piece of hardware. In Fast Ethernet, it is called a *media-independent interface* (MII). The MII links the MAC and PHY chips. It allows different types of PHYs to be controlled by one MAC. In 100 Mbps Ethernet, the interface is called the *media-independent interface* (MII) and in Gigabit Ethernet, it is called the *gigabit media independent interface* (GMII).
- The *PHY* encodes data from the MAC (e.g., Manchester or 4B/5B) and produces signal levels that can drive the magnetics and the cable.
- The *Magnetics* are an isolation transformer that protects the circuitry from voltage and current surges on the cable. They serve the same function that optical isolation serves in many other networks. Typical isolation is 1500 VDC but some industrial applications may require more.

Industrial-grade Ethernet hardware interfaces differ from office-grade gear in the following ways:

- The **common mode rejection ratio** is at least 40 dB and as high as 60 dB
- **Surge protection ratings are higher**, more than 2000 V instead of the standard 1500 V
- **More space between components** to prevent arcs
- **Transient protection circuitry** on transmit and receive sections

- **More copper on the circuit board** to reduce susceptibility to noise

Tip 5 – **When in doubt, take the lower-risk approach and select industrial-grade hardware.** The price difference is far less than "the real cost of not doing it right the first time."

3.4 Auto-Negotiation

Most NICs, hubs, and switches that support Fast Ethernet also support 10 Mb and automatically adjust their speed to match the node on the other end of the wire. This is tricky because there are seven possible Ethernet signals on an RJ-45 connector: 10BASE-T half or full duplex, 100BASE-TX half or full duplex, 100BASE-T2 half or full duplex, or 100BASE-T4.

While the details of Auto-Negotiation are beyond the scope of this book, it's important to realize that this handy feature saves you *lots* of time. Without it, you would be forced to shuttle back and forth between nodes, making manual adjustments until each node was in agreement. Given the number of nodes and potential distances involved, this alone would seriously dampen the world's enthusiasm for Ethernet.

Auto-Negotiation logic is incorporated in nearly all equipment shipped after 1996. Auto-negotiation is an upgrade of 10BASE-T link integrity and is backward-compatible with it.

3.5 Network Collisions and Arbitration: An Analogy

Imagine that you are having dinner with five other people, but the dinner table is 1 km wide instead of regular size. Assume

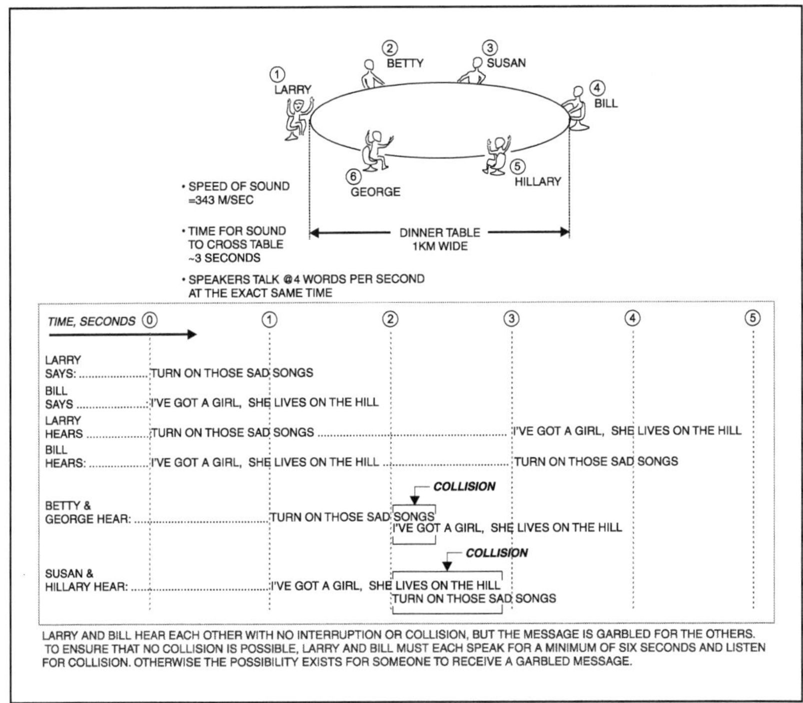

Figure 3-1. Collision Domain Analogy

for this illustration that you can easily hear each speaker despite the large distance.

Sound takes 3 seconds to travel 1 km. So if you and your friend across the 1 km table both start speaking at the same time, it will take 3 seconds before you know you are interrupting each other.

Successful contention rules would require the following conditions to be met:

- The rules that determine retransmit times must provide for at least 3 seconds of spacing between permitted transmissions.

- It will take each speaker at least 6 seconds to be certain that he or she is not being interrupted—if you start talking, your voice takes 3 seconds to reach the other side. Assume the other guy starts talking 2.99 seconds after you did. It will now take 3 seconds for his voice to reach you. That means you have to listen out for the total "Round Trip Time" of 6 seconds.

- Therefore *each speaker must talk for more than 6 seconds every time he or she has something to say*. If two people simultaneously talked for only 2 seconds each, they would each hear the other's message clearly. The other speaker's message would arrive 1 second after he finished speaking and he could hear it, but others around the table would hear both messages mixed together.

- The larger the table is, the longer the messages must be if everyone has equal opportunity to talk.

- In this example, message length was described in seconds, not bits or bytes. There is a direct relationship between allowable network length and minimum message length. At this table, if people speak at a rate of 4 words per second, then the minimum message size is 24 words.

- Suppose the baud rate goes up—extremely talkative speakers appear, who speak 40 words per second instead of 4. Then the minimum message size is now 240 words! When you move from 10 Mb Ethernet to 100 Mb or 1 Gb, the minimum required message length grows. However to maintain compatibility, you cannot do this. So you have to reduce the size of the table. So from 10 Mbps to 100 Mbps the frame stayed at 64 bytes min, so the collision domain shrank from 2,500 m (51.2 μs) to 250 m (5.12 μs).

In Ethernet, a message must be long enough to reach the other end of the network before the transmitter stops transmitting. The minimum message length defines a maximum network length, which is called the *collision domain.*

3.6 How the CSMA/CD Protocol Works

Whenever you interrupted your sister at the dinner table, did your wise and all-knowing parents remind you that you have two ears but only one mouth? They were teaching you a basic principle of communication. Dinner conversation is a contention/collision detection mode of communication. When there is a lull in the conversation, someone who has something to say speaks and "has the floor."

In Ethernet terminology, he does not hear a carrier signal from anyone else, and thus takes control of the network. Others who wish to speak must listen for a gap (**C**arrier **S**ense) and multiple people have the opportunity to take the next turn (**M**ultiple **A**ccess). When there is silence and two people speak up at the same time, they both hear the interruption (**C**ollision **D**etection) and if they are polite, they will both stop speaking and wait their turn. One will choose to speak first and then she "has the floor."

This is exactly how Ethernet works in half-duplex mode. (CSMA/CD is *not* required in full-duplex mode.) Each node listens to the wire and if another node is transmitting, the other nodes remain silent until the channel is free (i.e., no carrier is sensed). When the bus is quiet, a node with data to transmit will send it.

It's quite possible that another node also has data to send, and it starts transmitting at the same time. A collision occurs, also detected by both nodes. They stop and choose a random number that indicates how long to wait to retransmit. The one with the lowest number retries first, and the message can now successfully be sent. The *back-off algorithm* for choosing this random number is designed to minimize collisions and retransmissions, even when many, many nodes (up to 1024) are involved. A 10-Mb system generates back-off delay values ranging from 51 μs to 53 ms.

Tip 6 – **Ways to Reduce Collisions:**

- Minimize the number of stations on a single collision domain. Switches and routers divide a network into multiple domains.
- Avoid mixing real-time data traffic with sporadic "bulk data" traffic. If a network is handling regular cycles of I/O data, transferring a 10-MB file over the same network will compromise performance.
- Minimize the length of each cable.
- When possible, put high-traffic nodes close to each other.
- Use hubs and repeaters with buffers ("store and forward").
- Beware of "plug and play" devices that may clog a network as they search for other devices on the network.

Table 3-2. How an Ethernet Data Frame Is Constructed

Preamble (56 bits/7 bytes of 10101010---); used by the receiver to synchronize with the transmitter before actual data is passed	Start Frame Delimiter (SFD): (8 bits/1 byte, 10101011); indicates commencement of address fields	MACID of source (48 bits/6 bytes): 3 octets with NIC block license number – designates manufacturer + 3 octet device identifier	MACID of destination (48 bits/6 bytes; same format as source address)	TAG *	Type/ Length of data field 2 bytes	Message 0–1500 bytes	PAD 0–46 bytes Random "filler" data kicks in if message length is less than 46 octets to ensure minimum frame size of 64 bytes	CRC 4 bytes Cyclic Redundancy Check ("Frame Check Sequence," 32 bits/4 bytes)
		Three addressing modes: Broadcast: FFFFFFFFFFFF Multicast: First bit = 1 Point to Point: First bit = 0						

* TAG field in newer Ethernet frames adds 4 bytes for priority levels of quality of service. This supports advanced streaming and real-time services like VoIP (Voice over Internet Protocol). Tag field includes 3 bits for priority and 11 bits for VLAN address. * This is an IEEE 802.3 frame, not an Ethernet V2 frame; there are subtle differences.

Notes: The shortest possible frame = 64 octets with a message length of 46 bytes, excluding SFD and preamble. Longest possible frame = 1518, excluding SFD and preamble. The receiver detects the existence of pad data on the basis of the value in the Length field. All higher-level protocols (i.e., TCP/IP and anything "riding on top") operate entirely within the message field. The 48-bit physical address is written in pairs of 12 hexadecimal digits, as 12 hex digits in pairs of 2.

3.7 The Basic "Ethernet Design Rules"

The "5-4-3-2" rule states that the maximum transmission path is composed of 5 segments linked by 4 repeaters; the segments can be made of, at most, 3 coax segments with station nodes and 2 link [10BASE-FL] segments with no nodes between. Exceeding these rules means that some (though not all) nodes will be unable to communicate with some other nodes. You should check your design to ensure that no node is separated from any other node by more intermediate devices than the table below indicates.

Table 3-3. Maximum Transmission Path Between Any Two Nodes

5 segments 4 repeaters 3 link segments 2 coax segments	OR	5 segments 4 repeaters 3 coax segments 2 link segments

Note: This table is a popular simplification of the actual 802.3 rules.

3.8 "Would Somebody Please Explain This 7-Layer Networking Model?" (Adapted from *Sensors Magazine,* July 2001, ©Advanstar)

Networks, and the information that travels on them, are most easily understood in layers. For many years the International Standard Organization/Open Systems Interconnection (ISO/OSI) model (Figure 3-2) has been used as a way to represent the many layers of information in a network, particularly the low-level transport mechanisms. From top to bottom, these are the layers and how these layers relate to your product design (Table "Layer 1").

Please note that most networks do not actually use all these layers, only some. For example, Ethernet and RS-232 are just physical layers—layer 1 only for RS-232 and layers 1 and 2 for Ethernet. TCP/IP is a protocol, not a network, and uses layers 3 and 4, regardless of whether layers 1 and 2 are a phone line, a wireless connection, or a 10BASE-T Ethernet cable.

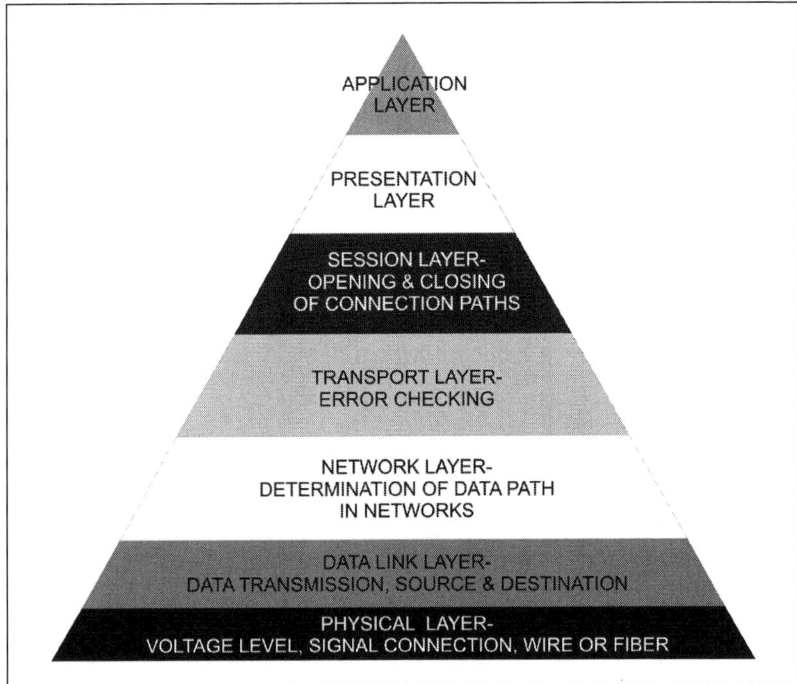

Figure 3-2. The 7-Layer Network Concept

Layer 7: Application

The application layer defines the meaning of the data itself. If you send me a .PDF file via email, the application that is used to open it is Adobe Acrobat. Many layers of protocols are involved, but the application is the final step in making the information usable.

In a sensor design, this is the software component that exchanges process data between the sensor elements (and their associated A/D converters, etc.) and the communications processor. It recognizes the meaning of analog and digital values, parameters, and strings.

J1939 and CANopen are application layers on top of CAN. FOUNDATION Fieldbus HSE is an application layer on top of Ethernet and TCP/IP. Modbus is an application layer on top of RS-232/485.

Layer 6: Presentation

The presentation layer converts local data into a designated form for sending and for converting received data back to the local representation. It might convert a character set such as MacRoman to ASCII for transmission. Encryption can happen in this layer.

Layer 6 is usually handled by application software and is not usually used in industrial networks.

Layer 5: Session

The session layer creates and maintains communication channels (sessions). Security and logging can be handled here.

Layer 5 is handled by software and is not commonly used in industrial networks.

Layer 4: Transport

The transport layer controls transmission by ensuring end-to-end data integrity and by establishing the message structure protocol. It performs error checking.

Layer 4 is usually handled in software (e.g., TCP/IP).

Layer 3: Network

The network layer routes data from node to node in the network by opening and maintaining an appropriate path. It may also split large messages into smaller packets to be reassembled at the receiving end.

Layer 3 is done in software.

Layer 2: Data Link

The data link layer handles the physical transmission of data between nodes. A packet of data (data frame) has a checksum, source, and a destination. This layer establishes physical connection between the local machine and the destination, using the interface particular to the local machine.

Layer 2 is almost always done in hardware with application-specific integrated circuits (ASICs). Low-speed networks can perform layer 2 functions in software.

Layer 1: Physical Layer

Layer 1 defines signal voltages and physical connections for sending bits across a physical media and includes opto-isolation, hubs, and repeaters. Physical media refers to the tangible physical material that transports a signal, whether copper wire, fiber, or wireless.

The data to be transferred starts out in the application layer and is passed down the seven layers to the physical layer, where it is sent to the receiving system. At that end, it is passed up through the layers to the remote application layer, where it is finally received by the user.

Just like a matryoshka[1] doll, when you encapsulate data at a particular layer, you must then wrap it in the lower layers from top to bottom; then when you unpack it, you must reverse the process.

Most protocols are related to the ISO/OSI model but do not follow the exact specification. Instead, they combine different layers as necessary.

3.9 Connectors

You don't have to spend much time investigating industrial Ethernet to discover that RJ-45 "telephone connectors" (Figure 3-3) are not viewed with a great deal of respect. Nor should they be. The design lacks even the most minimal environmental protection and can be easily damaged with a good yank on the cable. The surface area of the contacts is quite small and if the thin layer of gold over nickel is worn away by vibration, it becomes susceptible to corrosion and oxidation—not a great choice for your robotic welder, especially if downtime costs $15,000 per minute.

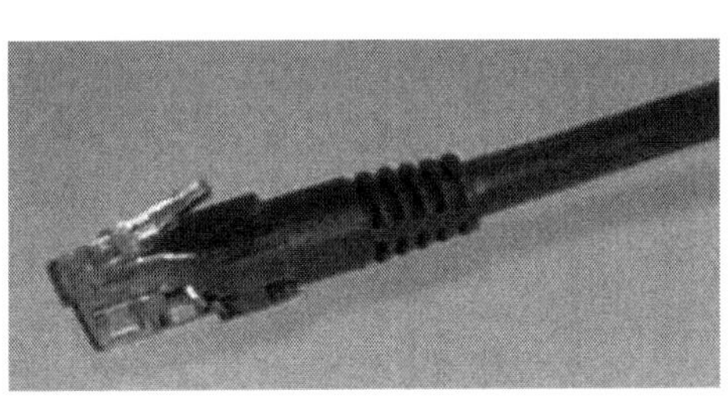

Figure 3-3. The Ubiquitous RJ-45 Connector for Ethernet

1. A matryoshka doll, also known as a Russian nesting doll or Russian doll, is a set of wooden dolls of decreasing size placed one inside another.

Tip 7 – **Fortunately there are alternatives—three in particular from the industrial world.** They have been designed to keep out liquids (e.g., IP65 or IP67), maximize contact surface area, and improve the sturdiness of the design. All of them facilitate feeding Ethernet cables through panels, simply by choosing appropriate receptacles.

IP67 Sealed Connector System for Industrial Ethernet

Many applications in medical, aerospace, food, and pharmaceuticals require absolutely secure and trouble-free communications. Other industrial applications are located in areas with extreme moisture, dirt, electromagnetic interference (EMI), and vibration. Standard office-grade RJ-45 connectors are not designed for these kinds of conditions. One of the ways to meet the requirements of these applications is to use sealed connector systems from vendors like Woodhead Industries, Phoenix Electronics, and others.

There are several other connector options that are being promoted for industrial applications. These options are less rigorous than the RJ-Lnxx Woodhead connector (see Figure 3-4) described above. For example, PROFINET recommends an IP20 version of the RJ-45 connector from Harting. Other vendors recommend M12 connectors. The connector choice for a particular application must be matched to the environment. In some applications, an office-grade RJ-45 will work just fine. In more dirty or dusty environments, an IP20 or M12 type connector might be best, while an IP67 type connector or sealed M12 connector (Figure 3-5) may be required for the most difficult environments.

Figure 3-4. Woodhead Industries RJ-Lnxx System

Figure 3-5. Sealed RJ-45 Connector

3.10 Pinouts

Table 3-4. RJ-45 Pinouts

Source: www.pin-outs.com

Pin No.	Function	Color
1	TX +	White/Orange
2	TX -	Orange
3	RX +	White/Green
4		Blue
5		White/Blue
6	RX -	Green
7		White/Brown
8		Brown

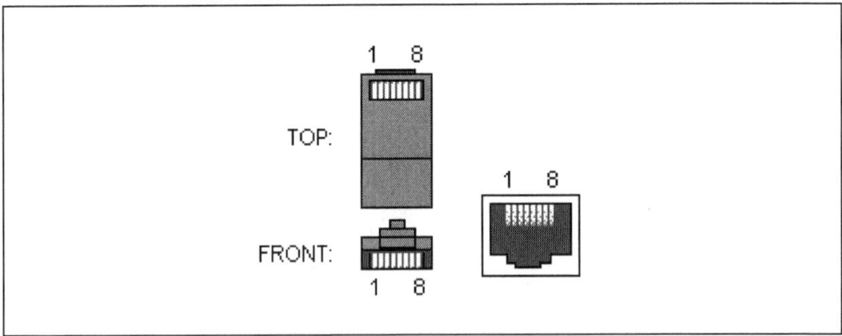

Figure 3-6. RJ-45 Standard Connector Pinout

The color scheme for crossover cable is used when a hub or switch port does not flip the transmit and receive pairs. It also can be used to link one PC NIC card to another. This ensures that receivers talk to transmitters and transmitters talk to receivers.

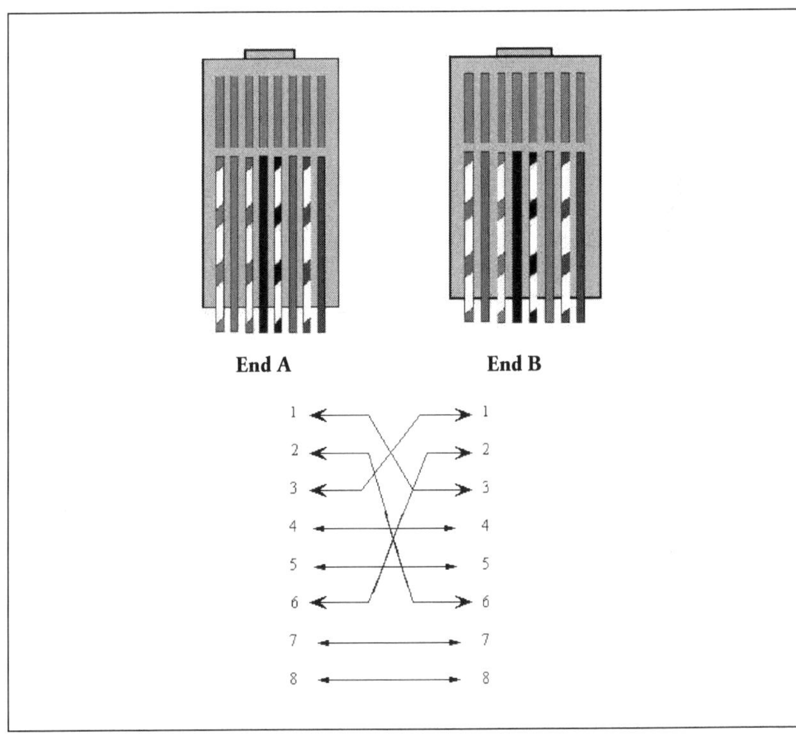

Figure 3-7. Color Scheme for Crossover Cable

Table 3-5. Category 5 Cabling Information (1000BASE-T Only)
1000BASE-T uses the standard registered-jack, RJ-45, connector.
Source: http://techsolutions.hp.com/dir_gigabit_external/training/techbrief6.html)

Connector RJ-45	Pin#	Description	ANSI/TIA/EIA-568A
	1	Transmit Data1 + (TxD1+)	white/green
	2	TxD1-	green/white
	3	Receive Data2 + (RxD2+)	white/orange
	4	RxD3+	blue/white
	5	RxD3-	white/blue
	6	RxD2-	orange/white
	7	TxD4+	white/brown
	8	TxD4-	brown/white

Ethernet DB-9 Connector

The trusty DB-9 connector (Table 3-5) has been employed for Ethernet systems, especially in Europe, and although few DB-9 designs are waterproof, it is certainly sturdier than an RJ-45.

Table 3-6. Pinout for Ethernet Using Standard DB-9 Connectors

Pin No.	Function	Color
1	RX+	White/Green
2	–	–
3	–	–
4	–	–
5	TX+	White/Orange
6	RX-	Green
7		
8		
9	TX-	Orange

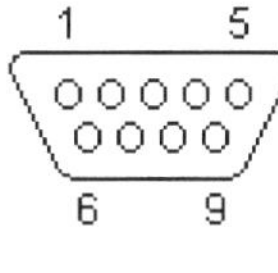

female view

M12 "Micro" Connector for Industrial Ethernet

This is based on the ever-popular 12 mm "micro"/"euro" design, which is popular in automation. It has eight poles and uses four of them for the Ethernet signal. The other four can presumably be used for other purposes (e.g., power *à la* IEEE 802.3af), as wise practice and future standards dictate.

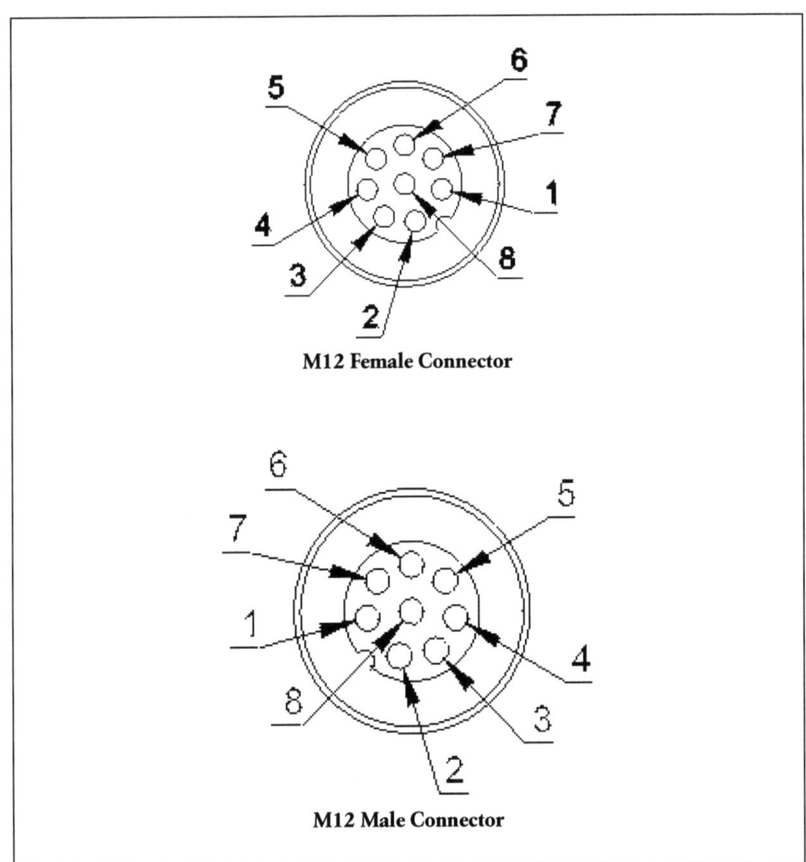

Figure 3-8. Male and Female Pinouts for Ethernet M12 Connectors

Pin No.	Function
1	–
2	–
3	–
4	TX–
5	RX+
6	TX+
7	–
8	RX–

4.0 Ethernet Protocol and Addressing

4.1 A Little Bit of History

Did you know that the networking cable connecting your computer to the Internet dates back to 1970? During that time period, a Harvard graduate student named Robert Metcalfe read a paper about something called *Aloha Net*. Aloha Net was a radio system used in the Hawaiian Islands to send small messages, also called *data packets*, between islands. A key feature of this network was that anyone could send messages at any time. If no acknowledgment was received, the message was not delivered and would be resent.

Dr. Metcalfe reasoned that with some mathematical enhancements to the system, the efficiency of the Aloha Net network, which then hovered between 15% and 20%, could be vastly increased. Not only did the efficiency increase all the way up to 90%, but the packet communications network he designed became the worldwide standard we know today as Ethernet (Figure 4-1). Now known as the IEEE 802.3 standard, the network still retains the elegance and simplicity of the original Aloha Net.

Dr. Metcalfe later went on to found 3Com Corporation, one of the leading manufacturers of Ethernet adapter cards and a major communications company. His discovery spawned billions of dollars in global wealth. Today Ethernet continues to gain momentum as more than half of the world's computers are linked to an Ethernet network.

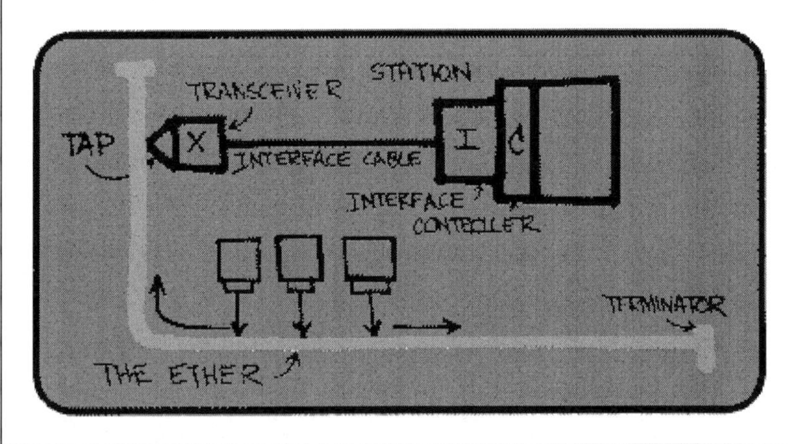

Figure 4-1. A Replica of the Diagram Drawn by Dr. Robert M. Metcalfe in 1976

4.2 The Ethernet Packet and How Messages Flow on Ethernet

A somewhat simplified diagram of the contents of an Ethernet packet is shown in Figure 4-2. This packet contains two address fields, some data, and a field that verifies correct reception of the packet.

Source Address	Destination Address	Packet Data	Error Checking Data

Figure 4-2. A Simplified View of an Ethernet Packet

The first two fields are unique 48-bit addresses of the sending computer and the destination computer. These addresses are not the familiar "192.168.0.10" type addresses we often see but are addresses assigned by the manufacturers of the physical Ethernet cards. Every manufacturer producing Ethernet hardware is assigned a series of 48-bit addresses. Known as the media access control (MAC) address, the manufacturers of Ethernet interface cards must ensure that they use only the addresses in their series and use it only once. That way, no two computers in the world can be assigned the same address. This is much more rigorous than the addressing system used by the post office. In the post office there can be many destinations with identical addresses. There can be a "1118" on Main street, one on Elm, and many others in other cities. How these 48-bit addresses are translated to the familiar dotted-decimal addresses is a subject covered a bit later in this chapter.

You should not make the mistake of confusing the "destination" 48-bit address in an Ethernet packet with the final destination of your message. The source and destination in an Ethernet packet are simply the addresses of the sending and receiving computers in the chain of computers between your desk and your best friend 2,000 miles away. It is much like taking a bus to work. In the middle of any two stops, the "source" address is the stop you just left, while your current "destination" is the next stop. Once you reach that stop, you can continue on to the next stop and continue with a new "source" and "destination," get off the bus and take a different one or begin work if you have arrived at your final destination.

In the same fashion, Ethernet packets simply flow from one computer to the next where one of three things can happen to the data field. First, it can be retransmitted to the next computer in the chain, which is akin to staying on the bus from our previous example. Second, it can be consumed by this computer if the final destination is this computer, or third, it can be discarded.

A message can be discarded for any number of reasons. It may be discarded if there is a checksum error on one of the several checksums in the message, if the message is not delivered in a timely fashion, or if the last field in the packet, the Error Checking field, indicates an error. Ethernet uses something called a Cyclic Redundancy Check (CRC) to verify accurate reception of a message. If the CRC is invalid, the message is rejected by your Ethernet interface and is never delivered to any of the TCP/IP protocol software on the receiving computer. In some cases, as we will see later, the sending protocol may expect a response and retransmit the message when the response is not received, or it may just ignore the lost message.

The CRC algorithm is an accurate method to use to detect message errors. Statistically, it can only misinterpret messages a few times in every 10,000 messages.

The firmware in your Ethernet interface hardware that checks the destination address, controls access to the network, and verifies the CRC is known as the MAC (media access control). This layer is not part of the TCP/IP protocol stack in your computer. Instead, the MAC software is a firmware layer that is part of the hardware interface to Ethernet. In the past, it was typically a stand-alone computer chip that you could identify by looking at your Ethernet hardware. In many cases, the MAC chip is now included as part of the intelligence on board your Ethernet interface hardware.

One of the tasks for the MAC software is to monitor the network and insert a message on the network when no other messages are detected. This operation is quite similar to driving and entering a highway from an on-ramp. Your car must fit into an open slot on the highway and if there is not an open slot, you have to wait for one. Unlike the highway, however, there can be thousands if not millions of crashes. When a crash, or collision, happens the MAC software hears the "crunch" as the bits collide and reschedules the message for a random time in the future. Hopefully, there is an open slot at that time. On busy networks, just as on busy highways, waiting for an open slot can delay your trip considerably. The process of detecting message collisions, rescheduling messages for a random time in the future, and retransmitting them is known as CSMA/CD or Carrier Sense Multiple Access with Collision Detection.

4.3 What Is the TCP/IP Protocol Suite?

If the destination address of a message matches the 48-bit address of your computer, the MAC software passes the message to the TCP/IP protocol suite. The TCP/IP protocol suite is a series of software programs that successively peel and process data packets. Each software program processes the data field remaining from being processed by the previous software layer. This is illustrated in Figure 4-3.

Protocols in this suite work together by passing their messages up and down the protocol stack. For example, the TCP protocol (described later in this chapter) takes application data and embeds it in the data field of a TCP message. It then passes the TCP message to the IP protocol where the entire TCP message becomes the data packet of the IP message. At each layer of the TCP/IP protocol suite, the software layer is only concerned with its fields and not the contents provided by previous layers.

The TCP/IP protocol suite is software and is usually a component of your computer's operating system. In Microsoft Windows®, Microsoft includes a TCP/IP protocol suite to process

Ethernet messages. In industrial devices, the TCP/IP protocol suite may be included as part of an operating system or may be a separate software component. The number of programs included in the suite is dependent on the vendor but always includes both the Internet Protocol (IP) and Transmission Control Protocol (TCP).

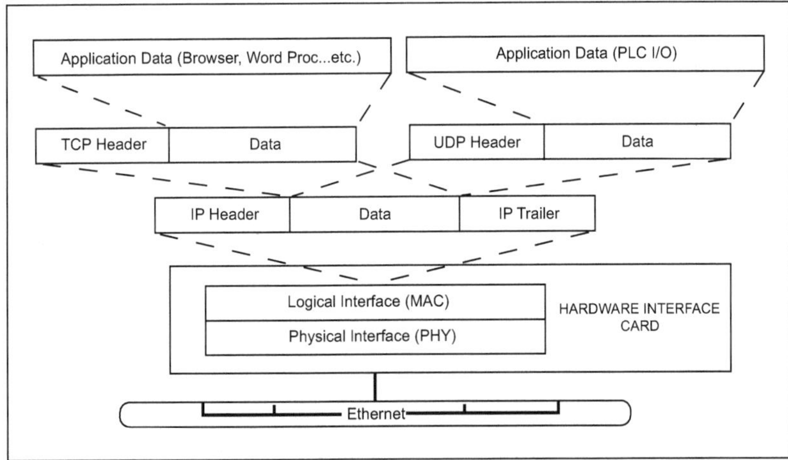

Figure 4-3. Data Flow Through the Hardware, IP Layer, and UDP/TCP Layers to Applications

4.4 TCP/IP Protocol Suite – IP Protocol

The IP part of the Internet Protocol Suite is the "Internet Protocol." It is used for almost all Internet communication. When a host sends a packet, it figures out how to get the packet to its destination; when receiving packets, it figures out where they belong. Because it does not worry about whether packets get to where they are going, nor whether they arrive in the order sent, its job is greatly simplified. If a packet arrives with any problems (e.g., corruption), IP silently discards it.

In addition to inter-network routing, IP provides error reporting and fragmentation and reassembly of packets for transmission over networks with different maximum data unit sizes. IP provides several services:

- **Addressing** – IP headers contain 32-bit addresses that identify the sending and receiving hosts. These addresses are used by intermediate routers to select a path through the network for the packet.
- **Fragmentation** – IP packets may be split, or fragmented, into smaller packets. This permits a large packet to travel across a network, which can only handle smaller packets. IP fragments and reassembles packets transparently.
- **Packet Timeouts** – Each IP packet contains a Time-To-Live (TTL) field, which is decremented every time a router handles the packet. If TTL reaches zero, the packet is discarded, preventing packets from running in circles forever and flooding a network.
- **Type of Service** – IP supports traffic prioritization by allowing packets to be labeled with an abstract type of service.
- **Options** – IP provides several optional features, allowing a packet's sender to set requirements on the path it takes through the network (source routing), trace the route a packet takes (record route), and label packets with security features.

The IP protocol makes great use of another protocol, the Address Resolution Protocol (ARP), to identify the specific destination of a message. When the IP protocol has a message to send, it checks its internal records to determine if it has the physical MAC address for the destination IP address. If not, it issues an ARP request to obtain the physical address. When the node with that IP address responds, it reports the MAC address that the IP protocol then uses to send the original message.

In addition to ARP requests that discover MAC addresses, there is also a "Reverse ARP" request, which discovers the IP address of a MAC address.

An interesting fact regarding the operation of the IP software is that it is "stateless" and operates silently (see Table 4-1). Stateless means that it does not remember or care about any previous operation. Unlike almost every other protocol in the TCP/IP protocol suite, each message is totally independent of all previous messages. Silent operation means that any invalidly formed packets or packets that can not be verified are simply discarded. No effort is made by the IP protocol software to inform the protocol layer above it that the message it encoded and sent never arrived.

Table 4-1. IP Data Packet

<table>
<tr><td>Version
4 bits</td><td>Internet header length
4 bits</td><td>Type of service
8 bits</td><td>Total packet length
16 bits</td></tr>
<tr><td colspan="2">Identifier (a pseudo-random tracking number)
16 bits</td><td>Flags
3 bits</td><td>Fragment offset
13 bits</td></tr>
<tr><td>Time-to-Live counter (255 max)
8 bits</td><td colspan="2">Protocol residing above IP
8 bits</td><td>Checksum header
16 bits</td></tr>
<tr><td colspan="4">Source IP address
32 bits</td></tr>
<tr><td colspan="4">Destination IP address
32 bits</td></tr>
<tr><td colspan="4">Padding and options</td></tr>
<tr><td colspan="4">Data (total packet length dictated by physical media;
for Ethernet, 1476 bytes)</td></tr>
</table>

Why IP Addresses Are Necessary

We discussed earlier that an Ethernet MAC address specifies a particular piece of hardware or a card in a computer. An IP address, on the other hand, is more portable and designates a "virtual" entity. A single IP address can represent one computer or a whole network of computers. Its virtual nature makes it portable and reassignable. Simply, a MAC address is like your name or social security number while an IP address is like your mailing address.

Ethernet nodes already have a MAC address; the MAC address could be compared with my vehicle identification number (VIN), which designates a unique piece of hardware. Now if all you had was my VIN number, it might still be difficult to find *me*. Someone else could be driving my car; I might have several cars myself. An IP address is like my mailing address, which is more helpful because it points you directly to me.

The Address Resolution Protocol (ARP) is the functional equivalent of calling the Department of Motor Vehicles and matching my VIN number to my postal address. It obtains the MAC address when the IP address is known. Reverse Address Resolution Protocol (RARP) is the exact opposite: Given a MAC address, it tells you the IP address. These two protocols allow the IP software to build the table of IP addresses and MAC addresses that are necessary to routing your message to its destination.

IPv4 addresses, the kind we now use, are 32 bits long, but writing the individual bits out is tedious. Instead, IP addresses are split into 4 separate bytes, each having 8 bits, represented by dotted decimals. For example:

204.101.19.6 = 11001100 01100101 00010011 00000110

IPv4 addresses are in short supply, and a new IP protocol, version 6, will eventually replace IPv4.

IP addresses are distributed by the Internet Assigned Numbers Authority (IANA); in turn, assignment of IP addresses is administrated by three sub-organizations: in Asia, www.apnic.net; in the Americas, www.arin.net; and in Europe, www.ripe.net. Internet Service Providers further sub-distribute IP addresses to end users.

The New Internet Protocol Version 6

IPv6 is an expansion of IPv4 and is designed to overcome a number of version 4 limitations. The most obvious improvement is the change from 32-bit IP addresses to 128-bit. This is necessary because the world is literally running out of IP addresses, which are carefully allocated. Other features include:

- More levels of addressing hierarchy
- Easier autoconfiguration of addresses
- Quality of service messages can be designated for high-priority routing through a LAN or the Internet
- Control over the route through which packets flow through a network via any casting
- Unicast (one-to-one) and multicast (one-to-many) messaging

IPv6 is only beginning to be implemented and will take some time to catch on.

Network ID versus Host ID

IP addresses are split into two parts: the NetID (a global designation, indicating a specific network somewhere on the Internet) and the HostID (a local address within a network, designating a specific machine). Example: in IP address 196.101.101.4, the HostID is 4 just as you read it. If the last digit is 0, it refers to "this network" and the NetID is 196.101.101.0.

Table 4-2. Drawing a Line Between the Network and the Host

Class	Leading bits	1st Byte	2nd Byte	3rd Byte	4th Byte	Range of Host
A	0	Network	Host			1.0.0.0 to 127.255.255.255
B	10	Network		Host		128.0.0.0 to 191.255.255.255
C	110	Network			Host	192.0.0.0 to 223.255.255.255
D	1110	Multicast Address				224.0.0.0 to 239.255.255.255
E	11110	Reserved for future use				240.0.0.0 to 247.255.255.255

Legacy Address Classes

To facilitate the parsing of IP addresses, five classes were defined long ago, ranging from a few networks with many hosts, to many networks with few hosts, as well as multicasting.

Today: Classless Subnet Masks

Class A, B, and C designations were used until it became apparent that large blocks of IP addresses were being wasted. Whenever a block of IP addresses is issued today, it is issued with a matching subnet mask. Specific notation is used for this:

An address of 203.14.4.13 with a mask of 11111111 11111111 11111111 11100000 (twenty-seven 1s and five 0s, or 255.255.255.224) is said to have a *prefix* of 27 and is written as 203.14.4.13/27.

Table 4-3. IP Address Classes, Networks, and Hosts

Class (designated by header bits)	Standard notation	NetID starts at bit #	NetID length (bits)	Number of possible networks	HostID starts at bit #	HostID length (bits)	Number of possible hosts per network	Netmask
A (0)	1.x.y.z to 126.x.y.z	1	7	126	8	24	16777216	255.0.0.0
B (10)	128.x.y.z to 191.x.y.z	2	14	16384	16	16	65534	255.255.0.0
C (110)	192.x.y.z to 223.x.y.z	3	21	2097152	24	8	254	255.255.255.0
D (1110)		Multicast address						
E (11110)		Reserved for future use						

Notes:

1. Address classes are a "legacy" system, which must be explained and understood. *However, today newly issued IP addresses are classless.*
2. 0.x.y.z is not allowed.
3. 127.x.y.z is reserved for loop-back testing, a simple self-test for proper configuration.
4. HostID = 0000000…. (all zeros) means "this network."
5. HostID = 1111111…. (all ones) means "All hosts on this network."
6. A subnet mask is a number that strips the HostID off of an address using the .AND. operation, that is, IP Address .AND. subnet mask = IP address with NetID only. This makes it easy to determine whether an incoming packet is destined for a particular local network. You must know the subnet mask for a given IP address to separate the HostID from NetID.
7. The "A" class theoretically also includes 127.x.y.z, so the number of machines is 1677214. A handy subnet calculator is available on the web: http://www.tcpipprimer.com/subnet.cfm

Assigning IP Addresses: Will Your Private LAN be Connected to the Internet?

Tip 8 – **You can use any IP address you wish so long as your LAN is never connected to the Internet; however if this network is connected to the Internet, there will be duplicate IP address conflicts.** The IP address police may hunt you down and arrest you. You then have two options:

1. Obtain unique IP addresses from your Internet service provider.
2. Use IP addresses that are *reserved* for private networks. Packets with these address ranges are not forwarded by Internet routers:

Table 4-4. Reserved Addresses for Private Networks

Class	IP address range	Number of possible combinations
A	10.0.0.0–10.255.255.255	16777216
B	172.16.0.0–172.31.255.255	65536
C	192.168.0.0–192.168.255.255	65536

Tip 9 – **If you use reserved IP addresses, then the gateway [router] between the LAN and the Internet must be configured as a proxy server to forward Internet packets to each reserved-address device.** Firewalls using network address translation can also do the trick.

***Tip 10* – Do not confuse the *reserved-address classes* A, B, and C here with the *general* IP-address classes A, B, and C, which were discussed a few pages ago.**

Reducing the Number of Addresses Routers Must Advertise with "Supermasks"

The speed with which you can locate sites on the Internet is because thousands of routers have "learned" where various IP addresses can be found. On the Internet, routers must "advertise" to other routers which IP addresses they serve. A mechanism has been devised by which a router can advertise large blocks of IP addresses with a single designation instead of thousands of separate addresses. This is called *classless inter-domain routing* (CIDR).

A subnet mask concerns the bits that create the NetID. A CIDR is different kind of mask; it concerns the bits in the IP address that are common to all hosts served by that router. So for a block of IP addresses, instead of advertising 2^N addresses, the router must advertise one address, one subnet mask, and one CIDR mask.

Example: A router serves 8 subnets, each containing 256 IP addresses. (See Table 4-5.)

Route advertisement: 161.200.0.0

Subnet mask: 255.255.255.0 or 11111111 11111111 11111111 00000000

CIDR mask: 255.255.248.0 or 11111111 11111111 11111000 00000000

Table 4-5. A Router Serves 8 Subnets, Each Containing 256 IP Addresses

Subnet	Range of IP addresses
1	161.200.0.1-254
2	161.200.1.1-254
3	161.200.2.1-254
4	161.200.3.1-254
5	161.200.4.1-254
6	161.200.5.1-254
7	161.200.6.1-254
8	161.200.7.1-254

Without CIDR, this router would have to advertise 2,048 IP addresses. With CIDR, only the route advertisement, the subnet, and the CIDR must be advertised by the router and stored by other routers.

A *default gateway* is an entry in a network configuration table that tells a device where to send a packet if the destination is outside the sender's subnet. The default gateway must be on the same subnet as the sender of the packet.

4.5 TCP/IP Protocol Suite – TCP Protocol

Transmission Control Protocol (TCP) is a protocol that *reliably* transfers messages between two computers. Where the IP protocol is concerned only about moving an Ethernet packet to the next node, TCP is concerned with providing a guarantee to the protocol layers above it so that it can with absolute certainty move data between one specific computer and another computer. In postal terms, this is "guaranteed receipt" mail. An indication is returned to show that the message arrived completely and correctly at the requested destination.

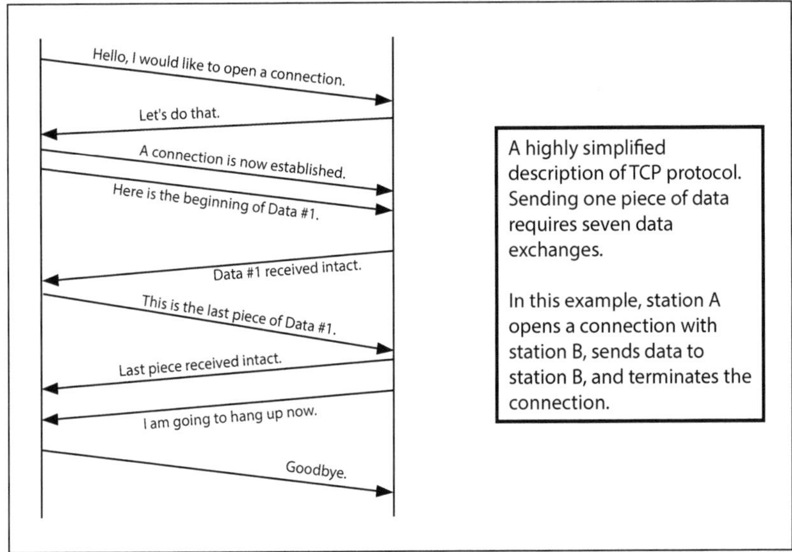

Figure 4-4. Handshakes in Opening and Closing a TCP Connection

TCP uses an inter-network to provide end-to-end, reliable, connection-oriented packet delivery. It does this by sequencing transmitted bytes with a forwarding acknowledgment number that indicates to the receiver the next byte the source expects to receive. Packets not acknowledged within a specified time period are retransmitted. This mechanism allows devices to deal with lost, delayed, duplicate, or misread packets.

TCP also offers efficient flow control, full-duplex operation (sending and receiving at the same time), and multiplexing. Multiplexing provides the capability to transfer numerous message streams over a single connection.

TCP provides the following set of services:

- **Stream Data Transfer** – From the application's point of view, TCP moves a continuous stream of bytes through the network or the Internet. The application does not have to parse the data. TCP groups the bytes in segments that are passed to the IP layer for transmission to the destination. TCP segments the data according to its own priorities.

- **Push Function** – Sometimes the application must guarantee that the data reaches its destination. So it pushes all remaining TCP segments in the queue to the destination host.
- **Close Connection** – Similarly, the application pushes the remaining data to the destination.
- **Reliability** – TCP assigns a sequence number to each byte and expects an acknowledgment (ACK) from the receiving station. If the ACK is not received within the timeout period, the data is sent again. Only the sequence number of the first data byte in the segment needs to be sent to the destination. There is no guarantee that packets will arrive in the exact order they were sent, so the receiving TCP puts the segments back in order on the basis of sequence numbers, and eliminates duplicate segments.
- **Flow Control** – When acknowledging receipt of a packet, the receiver also tells the sender how many more bytes it can receive without causing an overflow. This is designated by the highest sequence number it can receive without problems. This is also referred to as a window-mechanism.
- **Multiplexing** – Multiplexing is accomplished through the use of ports, just like User Datagram Protocol (UDP).
- **Logical Connections** – Reliability and flow control require TCP to initialize and maintain unique status information for each "conversation." The sockets, sequence numbers, and window sizes for this

conversation are called a logical connection. Every connection is identified by the unique pair of sockets used by the sending and receiving processes.

- **Full Duplex** – TCP can handle simultaneous data streams in both directions.

When you download large files from the Internet, you will observe some of these characteristics as connections are established.

- Downloads speed up and slow down with variations in network traffic.
- The server you are accessing adjusts its transmission rates to your connection speed.
- When you visit a new web page, pictures and graphics appear in sequence as new sockets are opened and closed.
- With each mouse click, data moves in both directions.

4.6 TCP/IP Protocol Suite – UDP Protocol

As an alternative to TCP, an application can choose to transmit a message to a destination using User Datagram Protocol (UDP). Unlike TCP, UDP is connectionless and does not provide an acknowledgment of receipt. UDP is the equivalent of first-class mail. Once you drop a letter in the mailbox, there is no record of transmission and no receipt of delivery.

UDP also does not guarantee message order. Ten messages sent over a UDP connection can arrive in an order different than the order transmitted.

UDP has much less overhead than TCP and can be thought of as "faster," in particular because it does not require acknowledgments. UDP is generally used for real-time applications like online streaming and gaming, where missing packets need not be resent. In these applications, damaged or missing packets are ignored as more recent data quickly replaces it. For example, if I am mailing you a daily letter describing how many tomatoes I have picked this season, there is no harm if one letter is lost. The next letter arrives with more recent information on the total harvest.

UDP is also employed where upper-layer protocols handle flow control and data stream checking and correcting, such as Netware and Microsoft Networking.

Common applications for UDP include:

- Simple Network Management Protocol (SNMP)
- Domain Name System (DNS)
- Trivial File Transfer Protocol (TFTP)
- Remote Procedure Call (RPC), which is used by the Network File System (NFS)
- Network Computing System (NCS)

4.7 Ports – How the TCP/IP Suite Is Shared Between Applications

The entire suite of TCP/IP protocols is shared between all the applications on your computer. You may have one window searching the Internet, another window running a spreadsheet on a remote server, and another window monitoring the stock market. Ports—virtual circuits for the transfer of specific data—are used to accomplish sharing of a TCP/IP stack.

TCP/IP defines 65,536 (2^{16}) available ports that are either common to all Ethernet users, registered to specific applications, or unassigned. Each standard service is assigned a port number. Here is a list of the most common services and their port numbers.

Port	Service	Purpose
21	FTP	File Transfer Protocol, great for downloading programs and files
23	Telnet	Allows remote configuration of a PC or smart device
25	SMTP	Used when *sending* email messages to a mail server
80	HTTP	Used to retrieve web pages
110	POP3	Used when *receiving* email messages from a mail server
139	NETBIOS	Used for file sharing in Microsoft networking
443	HTTPS	Used to retrieve secure web pages

A *port scanner* is a software program that probes your computer to detect open ports, which make the system vulnerable to security problems via unwanted external applications.

4.8 Other TCP/IP Application Layer Protocols

DHCP

DHCP stands for Dynamic Host Configuration Protocol and is a clever mechanism for temporarily, automatically assigning IP addresses in a network. It is used quite often.

A typical example of DHCP: You take your notebook computer on a customer visit, and they give you a desk to work at with an Ethernet cable. You configure your network manager in Windows to "obtain IP address automatically," and every time you boot up your machine, the local router gives it a temporary IP address. Now you can access the Internet and possibly share files with other PCs on that network every time you boot up. This is far easier than manually choosing an IP address (that someone else may have inadvertently taken) every time you go somewhere. Many LANs use DHCP for all their devices, simply for convenience.

When TCP/IP starts on a DHCP-enabled host, it sends a message requesting an IP address and subnet mask from a DHCP server. This server checks its internal database and then offers the requested information. It can also respond with a default gateway address, DNS address(es), or NetBIOS Name Server.

When the offer is accepted, it is given to the client for a specified period of time, called a *lease*. This process can fail if the DHCP server runs out of IP addresses.

SNMP

SNMP stands for Simple Network Management Protocol, and it allows monitoring and managing of a network. A device (automation product, PC, router, switch, or hub) must be enabled with an SNMP agent. The agent stores all variables related to its operation in a database called the Management Information Base (MIB).

The MIB defines all kinds of significant events (reboots, crashes, uplinks, downlinks) and reports such events.

TFTP

Trivial File Transfer Protocol (TFTP) is an extremely simple protocol to transfer files. It is implemented on UDP and lacks most of the features of FTP. It can read or write a file from or to a server. It is not secure and has no provisions for user authentication.

DNS

Domain Name System (DNS) is the convention used on the Internet to point domain names (www.yourcompany.com) to IP addresses. Since domain names are not likely to be very important in industrial networks, DNS will not be described in detail. However, its basic operation is similar in principle to Address Resolution Protocol (see ARP).

HTTP

HTTP stands for Hypertext Transfer Protocol, a TCP/IP protocol that enables the distribution of hypertext documents on intranets and the Internet. Just as with all other TCP/IP application layer protocols such as FTP and SMTP, HTTP is a client/server protocol.

The terms *HTTP server* and *web server* are somewhat interchangeable, although most web servers do far more than just HTTP. Similarly, *HTTP client* and *web browser* are roughly interchangeable, though most browsers do far more than just HTTP.

The web server is the application, sitting "above" the stack. HTTP is the application layer protocol in the upper layer of the stack, through which the web server accesses the stack in order to communicate with the client (like Internet Explorer or Google Chrome) on the remote machine.

FTP

FTP stands for File Transfer Protocol. It is very popular for moving files between computers. In an FTP session, two connections are opened. One is called a *control connection* and the other is a *data connection*; both use TCP. Each connection can have a different quality of service. Data transfer is always initiated by the client, but either the client or server can be the actual sender.

A popular freeware FTP utility is called WS_FTP and can be downloaded from many web sites. It conveniently moves, copies, deletes, and renames files between the local computer and a remote host.

Other popular FTP utilities: Coffee Cup FTP, Cute FTP.

Telnet

The Telnet utility provides a standard interface by which a client program on one host may access the resources of the server host as though the client were a local terminal connected directly to the server.

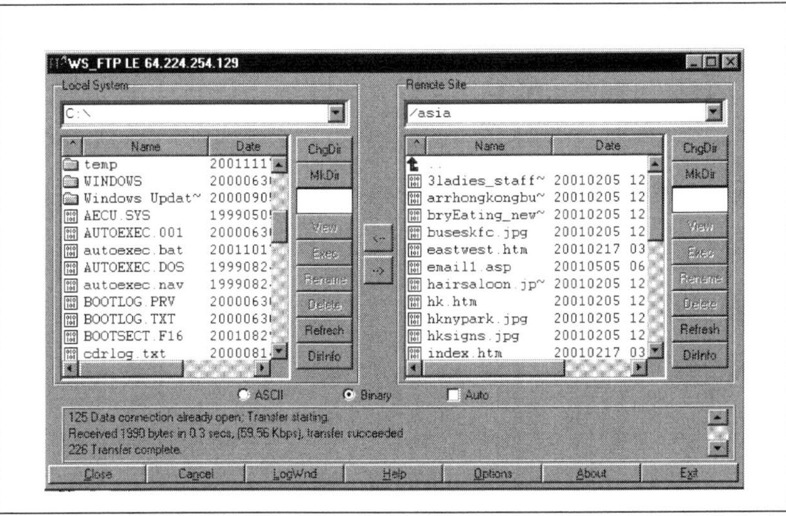

Figure 4-5. WS_FTP Is a Popular File Transfer Utility

A user on an Ethernet-connected workstation can talk to a host attached to the same network as though the workstation were a terminal attached directly to the host. Telnet can be used across WANs, the Internet, and LANs. Telnet allows the LAN-attached user to log in the same way as the local terminal user.

Most Telnet implementations do not include graphics features.

Telnet incorporates three concepts:

- A *Network Virtual Terminal* (NVT) is an imaginary device that applies a common data structure. Every host matches its own terminal characteristics to those of an NVT and expects every other host to do likewise.
- Telnet sessions use the same handshakes in both directions.
- Some hosts have more services than those supported by the NVT, so terminal features can be added and subtracted. Options may be negotiated, so client and server use a set of DO/DON'T/WILL/WON'T conventions to establish the characteristics of their Telnet session.

The two hosts begin by supporting a minimum level of NVT features. Then they negotiate to extend the capabilities of the NVT, according to the actual functions of the real hardware in use. Because of the symmetry in the Telnet sessions, both the server and the client may add options.

Tip 11 – **Many switches, smart devices, and Internet appliances, especially low-cost simple ones, use Telnet commands for configuration.**

4.9 Popular TCP/IP Utilities

PING

Packet InterNet Groper (PING) is the simplest TCP/IP utility, and one of the most useful utilities. It sends one or more IP packets to a destination host, requesting a reply and measuring the round-trip time.

Tip 12 – **The first test you should use to find out if a device is on a network is to attempt to ping it.** Normally if you can ping a host, then other applications like FTP and Telnet can also communicate with it. However with firewalls and other security mechanisms, which permit access to networks on the basis of application protocol and/or port number, this may not be possible.

How to use command line PING:

C:\> ping xxx.xxx.xxx.xxx (IP address) <enter>

or

C:\> ping www.nameofthesite.com <enter>

Tip 13 – **Especially when reading PING results on the Internet, remember that it is normal to have occasional packet loss. Also, PING times vary greatly and there is no single figure that designates a "problem" versus "no problem."** On average, the larger your bandwidth, the lower the PING time results. If you experience lousy PING results, the problem is related to the server being down or too busy to reply, or a router between you and the server is down or slow. You can use traceroute to test for this.

Figure 4-6 is an example of the PING utility executed from a DOS Window:

```
MS-DOS Prompt
Reply from 216.239.33.101: bytes=32 time=2003ms TTL=44
Reply from 216.239.33.101: bytes=32 time=2059ms TTL=44
Reply from 216.239.33.101: bytes=32 time=2064ms TTL=44
Reply from 216.239.33.101: bytes=32 time=2064ms TTL=44

Ping statistics for 216.239.33.101:
    Packets: Sent = 4, Received = 4, Lost = 0 (0% loss),
Approximate round trip times in milli-seconds:
    Minimum = 2003ms, Maximum =  2064ms, Average =  2047ms

C:\WINDOWS>ping www.google.com

Pinging www.google.com [216.239.33.101] with 32 bytes of data:

Reply from 216.239.33.101: bytes=32 time=2393ms TTL=44
Reply from 216.239.33.101: bytes=32 time=2059ms TTL=44
Reply from 216.239.33.101: bytes=32 time=1979ms TTL=44
Reply from 216.239.33.101: bytes=32 time=1944ms TTL=44

Ping statistics for 216.239.33.101:
    Packets: Sent = 4, Received = 4, Lost = 0 (0% loss),
Approximate round trip times in milli-seconds:
    Minimum = 1944ms, Maximum =  2393ms, Average =  2093ms

C:\WINDOWS>
```

Figure 4-6. Executing PING from a DOS Command Prompt

The first thing PING does is translate the URL into an IP address. It is then "pinged" with a packet of information 32 bytes long. After the packet is received, the server replies.

Time: This is the total response time for that particular packet.

Time To Live (TTL): This is the number of times this packet is allowed to be retransmitted by routers before being discarded. Each router that handles a packet subtracts one from this value. If TTL reaches zero, the packet has expired and is discarded.

Syntax: This is for DOS. The exact syntax varies with the OS:

> ping [-t] [-a] [-n count] [-l size] [-f] [-i TTL] [-v TOS] [-r count] [-s count] [[-j host-list] | [-k host-list]] [-w timeout] destination-list

Options:

-t	Ping the specified host until stopped.
-a	Resolve addresses to hostnames.
-n count	Number of echo requests to send.
-l size	Send buffer size.
-f	Set Don't Fragment flag in packet.
-i TTL	Time To Live.
-v TOS	Type Of Service.
-r count	Record route for count hops.
-s count	Timestamp for count hops.
-j host-list	Loose source route along host-list.
-k host-list	Strict source route along host-list.
-w timeout	Timeout in milliseconds to wait for each reply.

To see statistics and continue - type Control-Break;
To stop - type Control-C.

Netstat

Netstat asks TCP/IP the network status of the local host. Netstat reports:

- Active TCP connections at the local host
- The state of all TCP/IP servers on this local host and their sockets
- All devices and links being used by TCP/IP
- The IP routing tables in use

ARP

The ARP utility is especially useful for resolving duplicate IP addresses. A workstation is assigned an IP address from a DHCP server, but accidentally gets the same address as another workstation. You ping it and get no response. Your PC is attempting to find the MAC address, but it can't because two machines think they have the same IP address.

To resolve this problem, use the ARP utility to view your local ARP table and determine which TCP/IP address is resolved to which MAC address.

The ARP protocol belongs to TCP/IP, translating TCP/IP addresses to media access control (MAC) addresses with broadcasts.

In Windows, this table is stored in memory so that it doesn't have to do ARP lookups for frequently used TCP/IP addresses of servers and default gateways. Entries contain the IP address, the MAC address, and measurement of how long each entry stays in the ARP table.

The ARP table has two kinds of entries: static and dynamic.

Windows creates dynamic entries whenever the TCP/IP stack makes an ARP request and it can't find the MAC address in the ARP table. The ARP request is broadcast on the local segment. When the MAC address of the requested IP address is found, Windows adds that information to the table.

Static entries work the same way as dynamic, but you have to manually implement them using the ARP utility.

The ARP Utility

To use the ARP utility in Windows, follow these steps:

1. Open the cmd prompt window.
2. At the command prompt, type **ARP** plus any switches you need (see Table 4-6):

ARP -s inet_addr eth_adr [if_addr]
ARP -d inet_addr [if_addr]
ARP -a [inet_addr] [-N if_addr]

Table 4-6. ARP Utility Command Options

-a	Displays present ARP entries by examining the current protocol data. If inet_addr is specified, the IP and physical addresses for only the specified computer are shown. If more than one network interface uses ARP, entries for each ARP table are shown.
-g	Same as –a
inet_addr	Specifies an internet address.
-N if addr	Displays the ARP entries for the network interface specified by if_addr.
-d	Deletes the host specified by inet_addr.
-s	Adds the host and associates the Internet address inet_addr with the physical address eth_addr. The physical address is given as 6 hexadecimal bytes separated by hyphens. The entry is permanent.
eth_addr	Specifies a physical address
if_addr	If present, this specifies the Internet address of the interface whose address translation table should be modified. If not present, the first applicable interface will be used.

5.0 Basic Ethernet Building Blocks

Even a single Ethernet network can be quite extensive, with up to 1,024 nodes, hundreds of cables, and infinite possible combinations of hubs, switches, bridges, routers, network interface cards, and servers. This chapter describes these devices and their functions.

To understand the devices in this section the concept of a *collision domain* must be understood. Whenever two or more devices on an Ethernet segment begin transmitting at the same instant, there is a collision and neither message is transmitted. More and more collisions occur as the number of devices on a single segment increase until none of the messages can be delivered. Limiting the number of devices on a segment—the number of devices in a *collision domain*—solves this problem.

5.1 Devices

Hubs

Hubs are the simplest method of redistributing data on Ethernet (Figure 5-1). Hubs are "dumb," meaning that they do not interpret or sort messages that pass through them. A hub can be as simple as an electrical buffer with simple noise filtering; it isolates the impedances of multiple "spokes" in a star topology. Some hubs also have limited "store and forward" capability. In any case, hubs indiscriminately transmit data to all other devices connected to the hub. All of those devices are still on the same collision domain.

Note: Hubs are not assigned MAC addresses or IP addresses.

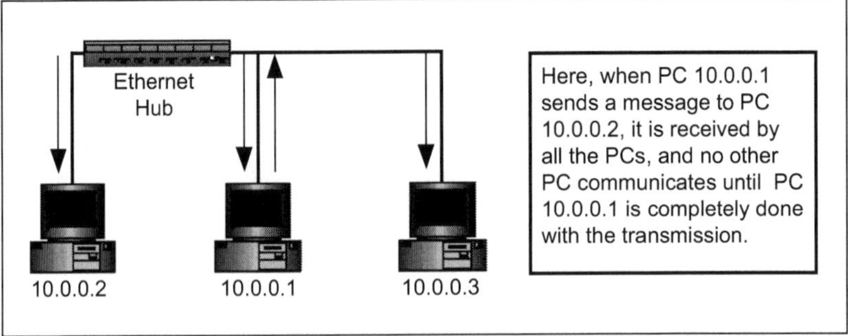

Figure 5-1. Ethernet Hub

Workgroup Hubs

Workgroup hubs are usually stand-alone units with four to eight ports. *Stackable hubs* often have many more ports and can be linked together to form a "super hub," which links even more devices to the same collision domain.

Segmented Hubs

Segmented hubs allow you to divide the available ports among multiple groups and collision domains. Each group you define is isolated from the others, as though you were using completely separate hubs. Bridges allow communication between segments.

Two-Speed Hubs

Two-speed hubs allow multiple baud rates to operate on the same hub, auto-detecting the data rate at each port and linking ports together with a speed-matching bridge.

Managed Hubs

Managed hubs have modest levels of intelligence and can be controlled remotely via a configuration port. This allows ports to be turned on and off, segments to be defined, and traffic to be monitored.

Repeaters

Repeaters are essentially two-port hubs. They simply clean up the signal and boost the signal level for large distances.

Stacking or "Crossover" Cables

Stacking or "crossover" cables allow multiple hubs to be daisy-chained. You cannot use standard cables to link two hubs together (or two NIC cards together) because they will link the transmit pins to transmit pins instead of connecting transmit pins to the receive pins. Some hubs have *crossover ports* that allow standard cables to be used.

Bridges

Bridges allow traffic to selectively pass between two segments of a network. Bridges operate at layer 2 of the OSI model and effectively extend the reach of each segment. Bridges make their forwarding decisions on the basis of the MAC address.

Tip 14 – **You should use a bridge when your network traffic can be clustered into "devices on segment A that mostly talk just to each other" and "devices on segment B that mostly talk just to each other."** The bridge handles the cases where devices on A must talk to B, and the rest of the time it reduces the traffic on each side.

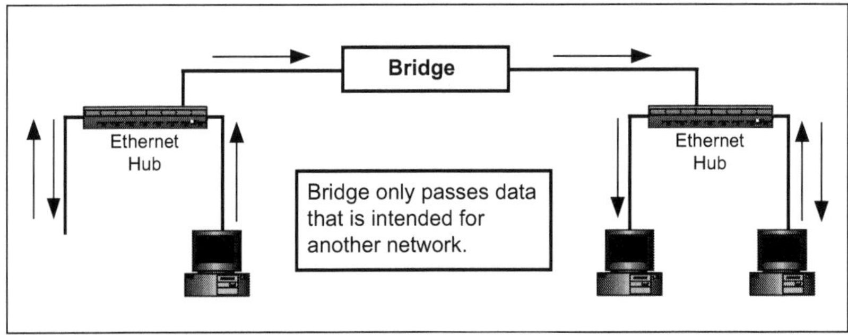

Figure 5-2. Bridge

Intelligent Bridges

Intelligent bridges (Figure 5-2) learn over time what devices are connected on each side and "figure out" which messages to forward and which ones to block. Such a bridge will automatically adapt to changes made to the networks over time.

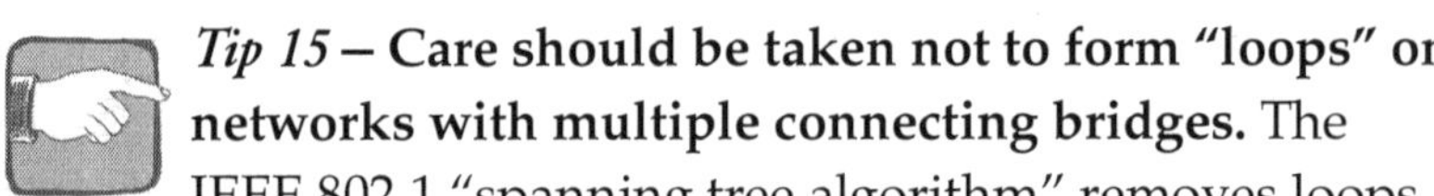

Tip 15 – **Care should be taken not to form "loops" on networks with multiple connecting bridges.** The IEEE 802.1 "spanning tree algorithm" removes loops. One bridge in a loop becomes the "root" and all other bridges send frames toward the root bridge.

Note: Bridges are not assigned MAC addresses or IP addresses.

Switches

A switch is an intelligent bridge with many ports. A switch learns what addresses are connected to each port and sends messages only to their intended destinations.

There are two basic types: *Cut-through switches* forward a message to their destination as soon as they recognize the intended address. *Store-and-forward switches* hold the packet in memory and examine the entire contents of the packet first. This enables it to trap errors and prevent those bad packets from being sent through the network. A store-and-forward switch can also hold the packet until traffic on that segment disappears, reducing collisions.

Switches usually operate at layer 2 (switching decisions based on Ethernet MAC address) and some operate on layer 3 (switching decisions based on IP address). Layer 3 switches can be used in place of routers.

Full-duplex switches handle both transmit and receive lines simultaneously.

Multispeed switches handle segments with multiple data rates. It is now common for switches to support multiple speeds on every port. With older models, there was some delay or "bottleneck" from aggregating the data but that is no longer the case.

Many industrial applications require a level of determinism, and Ethernet is often criticized for its lack of determinism. Switches make deterministic performance possible by eliminating collisions. IEEE 802.1p allows prioritization of messages at layer 2 with a 16-bit additional header. Rigid priorities on the network allow important messages to be sent without collisions.

Note: Switches are not assigned MAC addresses or IP addresses.

Routers

A router's job is to forward packets to their destination, using the most direct available path. Routers make their decisions on the basis of IP addresses (see Figure 5-3). When a packet comes in, a lookup table determines which segment it should be routed to.

Many times the segment has only another router instead of a final destination. The implication is that the intended destination is remote. The IP address of the next router (there could be many) is called the *default gateway.*

Routers maintain tables of IP addresses on each segment and "learn" the most direct paths for sending data. When the network changes, it takes time for routers to accommodate the new information. On the Internet, if you have a web site www.yourwebsite.com, the URL points to the IP address of the host server. If you change hosts, you get a new IP address and register the change with the Domain Name Server. It may take several days for DNS servers across the Internet to update their tables and point you to the new IP address.

Types of Routers

- **Two-Port Routers** – Two-port routers link only two networks.
- **Multi-Port Routers** – Multi-port routers link several networks.

- **Access Routers** – Access routers use modems (e.g., ISDN, v.34, v.90) to access the Internet, usually on-demand.
- **Bridging Router (Brouter)** – Bridging routers, or brouters, change from being routers to being bridges when they receive a packet they don't understand. They just go ahead and send the message.
- **Terminal Servers** – Terminal servers connect multiple serial devices (RS-232, -422, -485) to Ethernet. The term comes from terminals on mainframes, connected to the LAN via serial port.

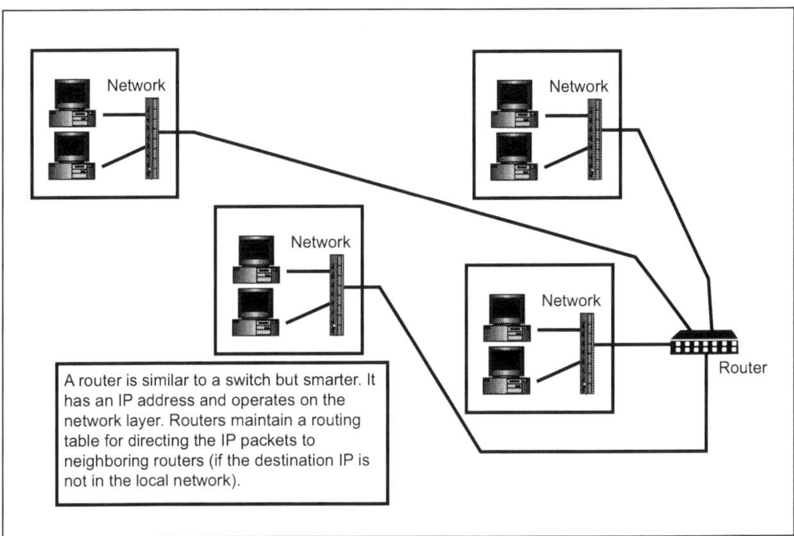

Figure 5-3. How a Router Works

- **Thin Servers** – Thin servers link a single device to Ethernet and allow COM ports on the other side of the network to appear as though they are local to your PC, even though they may be on the other side of the world.

- **Network Time Servers** – Network time servers use a Global Positioning System (GPS) to provide accurate local time for synchronization of devices and time stamping of events.

Gateways

Gateways convert messages from one protocol to another. Examples could include Modbus on RS-232 to Ethernet Modbus/TCP, or DeviceNet to EtherNet/IP. In most cases, the physical layers, protocols, and speeds are all different.

Tip 17 – **Gateways normally require configuration to work properly and are normally thought of as "band aids" rather than permanent or global solutions.**

Interface Cards

A NIC links your PC to Ethernet via the PCI, ISA, PCMCIA, PC/104, or other buses. NICs handle layers 1 and 2, while the host processor in the PC handles everything else.

5.2 Determinism, Repeatability, and Knowing if It's "Fast Enough"

Many people confuse determinism with speed. A definition of terms is helpful:

Deterministic means that a system is guaranteed to respond within a designated period of time: less, but not more. The term really is not that meaningful unless a time specification is included with it.

Example: "This PLC is deterministic to 10 milliseconds" means that when an input state changes, the corresponding output state change will occur no more than 10 milliseconds later.

The term *repeatable* defines the space between the lower *and* upper limit for response time. A spec for repeatability designates the width of a time window.

Example: "This PLC is repeatable within 2 milliseconds" means that the response time will never vary by more than ±1 milliseconds. This statement does not specify the response time though, only the *consistency* of the response.

***Tip 16* – A full understanding of a system's response capabilities normally requires specification of both determinism and repeatability.** Some automated processes require determinism and repeatability; others require only determinism; others require one or both only "most of the time"; and still others require neither.

Your watch is both deterministic and repeatable, within microseconds. An E-Stop must absolutely be deterministic, but repeatability per se is not usually so important—there's no good reason to have a fixed amount of delay before the system shuts down. A pneumatic valve that rejects black grains of rice into a waste bin on detection by a vision system must be deterministic *and* repeatable, otherwise the puff of air may come too soon, which is just as bad as coming too late.

"Fast enough" is another issue. Most processes do not require absolute determinism; they really require that systems respond within a certain amount of time, most of the time. So you define the requirement as a statistical probability that the system will respond within, say, 10 milliseconds, 99.9% of the time.

In those cases, even a nondeterministic collision-based Ethernet system will be acceptable, so long as the network loading is below acceptable limits 99.9% of the time.

Achieving Determinism on Ethernet

Half-duplex Ethernet has inherent collision problems, and though the CSMA/CD protocol offers a good solution, it is inherently nondeterministic. Complex formulas exist for calculating network loading and response-time probabilities, but *if true determinism is necessary, you should isolate collision domains by using switches instead of hubs.*

With switches, the remaining determinism problems are caused by their throughput limitations. If the switch is not able to handle the full speed on each port, or if the number of packets sent to an output port exceeds the bandwidth of that port and fills the output buffer, this causes a nondeterministic buffering delay. Higher protocol layers at the stations must handle lost packets.

The following methods are used to prevent switches from overloading:

- **Flow Control** – The switch sends PAUSE packets on a full-duplex port if the number of packets received on the port is more than the switch can handle.
- **Back Pressure** – If the traffic load exceeds the switch's capacity, the switch acts like a port operating in half-duplex mode. It makes the transmitter think the collision domain is busy.
- **Priority** – Ethernet packets that are designated as high priority are put in a high-priority queue. Those packets are sent ahead of the low-priority packets, which might possibly be dropped. This is the most "deterministic" approach to the problem.

How Priority Messaging Works

Many switches now support priority, with two or more output queues per port, where the higher priority queue(s) are designated for time-critical data with quality of service. Each vendor uses a different algorithm for selecting the queues, but in general there are two approaches:

- **Round-Robin** – After X packets are unloaded from the high-priority queue, a low-priority packet gets its turn to go out.
- **High Priority Always Takes Precedence** – Low-priority packets are sent only when the high-priority queue is empty.

In any case, a high-priority message can still be delayed by a low-priority message if the low-priority message has already started transmitting when the high-priority message enters the switch. But it's not that difficult to calculate worst-case scenarios for this.

The worst-case delay for high-priority packets holds, regardless of low priority traffic. Several priority implementations exist with respect to how a packet is identified as a high-priority packet. The priority handling depends on the switch functionality.

How Switches Determine Priority

- **Based on MAC Addresses** – Both the MAC source and destination address can be prioritized. The switch must be a "managed" switch so the user can choose high-priority MAC addresses. This is a fairly rigid approach.
- **Based on High-Priority Physical Port** – One or more switch ports can be designated as high priority, so all

packets received on these ports are considered high-priority packets. Most switches that work this way are "managed." A *managed switch* is one that is externally configured for optimum performance—this provides extra functionality with the disadvantage of extra complexity.

- **Based on Priority Tagging** – IEEE 802.1p and IEEE 802.1q designate an additional Tag Control Info (TCI) field for the Ethernet MAC header. This adds a 3-bit priority field that is used for priority handling, allowing eight levels of priority. Most priority tagging Ethernet switches have only two or four queues, so the network configuration must account for this limitation. The advantage is that no switch configuration is necessary. Unfortunately most stations today do not support priority tagging. The switch can be configured to remove the tags after switching, but this requires managed switch operation. Another problem could be that other switches do not support priority tagging and will not forward the longer packets.

Caution: *Even though they may be rated for 10, 100, or 1,000 Mbps operation, many switches cannot sustain full-traffic loads at their rated speeds. This can create problems on high-speed deterministic systems. If this presents a potential problem, ask your vendor what the throughput of their switch is, and how the switch allocates its time to competing devices.*

Drivers and Performance

All other network layers, including TCP/IP, are handled in software. Nearly all PC operating systems, including Windows, Linux, DOS, UNIX, VxWorks, etc., have TCP/IP built in. Software applications for control or operator interfaces have drivers that pass application data, including higher-layer protocols like Modbus/TCP and EtherNet/IP to TCP/IP.

Tip 18 – **There can be significant performance issues and delays with respect to driver and application performance, and there are no defined standards for driver performance.** Many drivers simply are not written to serve the needs of deterministic applications.

Response time can vary considerably on the basis of CPU speed, memory, how well the drivers are written, what other applications are running on the PC, etc.

Tip 19 – **There can be no doubt that in most cases, the speed of industrial Ethernet networks will be limited by software and drivers, and not by Ethernet itself.**

6.0

Network Health, Monitoring, and System Maintenance

By Mark Mullins. Reprinted with permission from the **Industrial Ethernet Book**, *www.ethernet.industrial-networking.com, ©2001 GGH Marketing Communications and Fluke Networks.*

6.1 What Is It that Makes a Network Run Well?

Fluke Networks has profiled dozens of networks worldwide in an effort to determine the answer. In our research, the best-run networks had thirty-five times less downtime, resulting in annual savings of over $227,000. Not surprisingly, users of these networks were the most satisfied of all groups studied. One surprising conclusion is that the number of support staff per end-user of these well-run networks was actually lower than that of the poorly-performing networks.

So how does a network support group enter this desirable group? In studying this question, Fluke Networks uncovered seven "best practices" of well-run networks. They are:

- Management involvement

- Preparation and planning
- Problem prevention
- Early problem detection
- Quick problem isolation and resolution
- Invest in tools and training
- Quality improvement approach

Having the right tools for monitoring, documenting, and troubleshooting your network helps with nearly all of these areas. Let's look at each of these three functions and discuss the tools for each.

Monitoring

Monitoring your network is essential in identifying problems before they become serious, and just to have a general idea of what is going on in your network. When monitoring, there are key questions you'll need the answer to, such as:

- Who's talking to whom?
- What are they talking about?
- Are there problems out there?

The most important tools are protocol analyzers, embedded RMON (Remote MONitoring) agents, and external RMON agents.

Protocol analyzers allow the network engineer to capture traffic passing by on the network, and then decode that traffic in order to understand the traffic. For example, a single frame may contain IP-addressing information, TCP flow control, and HTTP commands. The analyzer needs to be able to decode each of the protocols so that the user is not presented with a bunch of unintelligible hexadecimals. Billions of such frames may travel over a network in the course of a day, so the analyzer needs to be set up to filter, that is to capture only certain types or sources of frames, as well as trigger, or start capturing traffic after a certain type of frame is detected.

Low-cost analyzers have trouble keeping up with fast or busy networks, while higher-priced ones offer more memory and specialized hardware to keep up with even Gigabit Ethernet. Higher priced analyzers also decode more protocols and offer expert analysis of traffic to find problems faster.

Embedded agents are found in most of today's switches and routers to collect information on activities in each interface of the device. This information is stored in a management information base (MIB) and can be accessed by devices using Simple Network Management Protocol (SNMP). External RMON agents do much the same thing, except that they offer much greater depth of information, and they have to be purchased (where embedded RMON is usually a no-cost feature).

In terms of depth, embedded RMON agents generally offer only a very high-level overview of what's happening on the network interface: utilization statistics (how 'busy' the interface is) and error counts. Some embedded RMON agents can also generate alarms when certain thresholds are exceeded. External agents, on the other hand, generally offer much greater detail, such as which devices are using the port, and the ability to capture traffic, like a protocol analyzer.

Most newer external agents support RMON2, which adds the ability to track application-layer traffic. This is important in determining who is really using bandwidth. A quick comparison of the detail yielded by the three approaches is instructive.

An embedded RMON agent would tell you that Ethernet Interface 42 is very busy. An external RMON probe could tell you that a certain PC is sending a lot of traffic to the router. A RMON2 probe could tell you that 'Bob's PC' is sending HTTP traffic to the website hotjobs.com. Obviously, RMON2 provides much more detail. And, not surprisingly, that detail comes at a cost.

Monitoring Switched Networks

The main issue with external probes and protocol analyzers is where to put them. Before the advent of switched Ethernet, a probe placed anywhere on the network could see all the traffic on the network and provide total visibility. Today, switched networks provide higher performance and faster response, and are recommended for industrial networks. Unfortunately, switched networks only send traffic to the intended recipient, so special allowances must be made so that the probe can monitor the relevant part of the network.

One of the simplest methods is to use port mirroring. By using appropriate commands to the switch, it can be configured to copy traffic at one interface to another one where the probe is connected. The advantage of this method is the fact that it allows you to monitor whichever port you want. The disadvantage is that while monitoring that port, you have no idea what's happening on any other port. In addition, port mirroring will generally forward only good frames, and often, it's the bad ones you're looking for. Finally, if the switch is very busy it may not send all the frames to the mirror port—so that one critical frame causing the problem might be missed.

Tip 21 – **For more thorough monitoring, a tap may be installed into critical links.** This is simply a hardware device that allows the probe to see all the traffic on the

link. Unlike mirroring, taps never miss a frame or an error. However, they have to be installed on every port that must be monitored. And like mirroring, they give visibility into only one link at a time.

To get complete vision when monitoring a switched network, a combination of approaches must be used. Embedded agents can be monitored to get an overview of what's happening on every interface in the network. External RMON2 agents can be installed with taps to provide constant monitoring of key links, such as those between switches, to servers, and wide area networks. Additional RMON agents can be connected to switches and connected as needed using port mirroring to monitor problem ports found with the embedded agents.

Documenting

Tip 22 – **Documenting your network is essential for two reasons.** First, when it becomes necessary to upgrade or expand your network, you will need to know where to start. Second, knowing the normal state of your network is essential when it is time to troubleshoot. If a doctor didn't know the values of a normal temperature or blood pressure, then those measurements would be meaningless. Each network has its own normal operating condition, so it is important to know what yours looks like—before a problem arises.

A good example of this is a network at an automotive plant in Michigan, where Fluke Networks was offering training on network documenting. This customer had recently upgraded its network and was extremely pleased with how well it was operating. In the course of the class, we documented many interesting characteristics of this network: over 1,100 stations in one collision domain, over 400 errored frames in two hours, sustained peak utilization over 70%, and an average of 2% collisions. Any of these would be considered serious problems in most networks, but this was 'normal' for them. If they ever experience a problem they need only look for what changed, rather than waste time tracking down unrelated issues.

Documenting is the process of recording the state of your network. There are two main questions: What is out there? And, how is it performing? By keeping records provided through monitoring systems, a good idea of normal performance can be obtained. Other additional documenting tools can be valuable.

The first of these can be called SNMP ping monitors. These discover the devices in the network and then gather SNMP information from these devices. They also offer some powerful documenting features. For example, Microsoft Visio finds the devices and then provides a complete network diagram at a very reasonable price. Fluke Networks' Network Inspector provides a wide variety of reports, and can plot SNMP statistics for up to a 24-hour period. Other reporting packages are available from Concord Communications, InfoVista, and Visual Networks.

Troubleshooting

When something goes wrong, it is usually important that it get fixed as fast as possible. Many of the tools noted above can help with the troubleshooting process. In fact, if everyone installed everything noted above, and never let any users or applications near the network, there would not be much need for specialized troubleshooting tools. For the real world, however, specialized tools are available to solve problems fast.

The most common of these is the protocol analyzer, loaded onto a laptop computer. While in-depth analysis capabilities allows this to tackle the most challenging of network problems, the complex set-up and limited vision in switched networks make them a troubleshooting tool of last resort.

The next most common tool is the *cable tester*, which can help track down the most common cause of network problems—cabling. Basic and advanced types are available. Basic testers will find broken cables, shorts, and split pairs. A split pair occurs when two channels constituting a transmit plus and transmit minus are not connected to a pair that is twisted together. The result can be transmission errors or even a complete breakdown in communications. Many low-cost testers cannot find this problem—use of these is not recommended. Some basic testers can display the distance to the fault, which can greatly speed troubleshooting.

Tip 23 – **Advanced cable testers not only find these common problems, but can also determine the performance level of the cable.** In the days of 10 Mbps Ethernet, almost any cable could handle the requirements. Higher-speed networks, at 100 Mbps and gigabit speeds, place significantly higher demands on the cabling and more advanced tools are needed to determine if the cable is up to the task. These advanced testers measure parameters, such as crosstalk and return loss, and can certify the performance of cabling for high-speed networks. They also cost about four times what a basic tester costs! If cable performance is verified with an advanced tester at installation, most sites need only the basic tester for daily troubleshooting. However, as higher performance networks become more common, the need for advanced testers will grow accordingly.

A new class of tester, the Integrated Network Analyzer, offers the fastest approach to network troubleshooting. These devices incorporate the most commonly used capabilities of protocol analyzers and cable testers to provide a complete solution for troubleshooting. Portability and ease of operation are two key features. Some also offer advanced features such as the ability to discover devices on the network (like SNMP ping monitors) and SNMP queries of network devices.

__Mark Mullins__ is Marketing Manager for Enterprise Systems at Fluke Networks in Everett, Washington, USA. He holds a BS in Computer Science and an MBA from the University of Washington.

6.2 Popular PC-Based Ethernet Utilities, Software, and Tools

PC-based network sniffers operate on a PC that links to an Ethernet LAN via a hub (not a switch) and collect data as it goes by. They break down and organize Ethernet frames and/or TCP/IP packets so you can make sense of what's happening on your network. Use them to identify chattering nodes, corrupted data, and mysterious sources of network traffic.

If your TCP/IP sessions "hang up," a sniffer might tell you which device sent the last packet and which one failed to respond. Similarly, if devices are responding slowly, time stamps will show you which system is waiting and which system is responding slowly. The sniffer can monitor broadcast or multicast storms and packet errors. By recording and displaying the traffic on the Ethernet wire, or a filtered segment of the traffic, you will pinpoint problems and intelligently improve network performance.

There have been many network sniffers in the past but now one package dominates the market, Wireshark (see Table 6-1).

Table 6-1. Wireshark

Product Name	Wireshark
URL	https://www.wireshark.org/
Platforms Supported	Windows, Linux
Licensing	Open Source License
Description	Wireshark provides a very powerful set of features: • Capture live packet data from a network interface. • Open files containing packet data captured with tcpdump/WinDump, Wireshark, and several other packet capture programs. • Import packets from text files containing hex dumps of packet data. • Display packets with very detailed protocol information. • Save packet data captured. • Export some or all packets in several other capture file formats. • Filter packets on many criteria. • Search for packets on many criteria. • Colorize packet display based on filters. • Create various statistics.

7.0 Installation, Troubleshooting, and Maintenance Tips

7.1 Ethernet Grounding Rules

Tip 24 – **The shield conductor of each coaxial cable must be grounded at one point only; otherwise you will create ground loops.** On coax, this is often done at the location of a terminator. Many terminators provide screw terminals for this purpose. You should check for exposed wire at other locations since it could make contact with other conductors or ground points.

Ethernet Grounding Rules for Coaxial Cable

- **10BASE-5 (Thick Ethernet)** – Grounding is a requirement.
- **10BASE-2 (Thin Ethernet)** – You can ground if your local electrical code requires it.

Grounding coaxial cable is generally good; it dissipates static electricity and makes your network safer. Many local electrical codes require network cables to be grounded at some point.

Tip 25 – **Many Ethernet segments are not grounded though, and grounding can add complications to an otherwise working network.** But always follow the electrical codes. A segment should be grounded only at one end of the coaxial segment.

Tip 26 – **Do not use copper cables to link buildings!** Copper cable attracts lightning strikes.

The ground potential between the two buildings may be different. This can introduce transient voltages and any number of dangerous problems.

Tip 27 – **Use fiber to connect buildings instead.**

Twisted-Pair Cable Types

Twisted-pair cabling is categorized as follows:

- Category 1 is mostly used for telephone connections. Do not use for computer networking.
- Category 3 works up to 16 Mbps and may be the most commonly installed twisted-pair format.
- Category 5 works up to 100 Mbps and is the most popular kind of cable sold by computer vendors.
- Category 5E supports Gigabit Ethernet and is preferable if there is any possibility of upgrading your system to 1000 Mbps.

- Category 6 includes all the CAT5E parameters but extends the test frequency out to 250 MHz, exceeding current Category 5 requirements. Category 6 will be the most demanding standard for four-pair UTP terminations based on RJ-45 connectors.

Tip 28 – **When you use a specific grade of cable, all components and interconnections on the network must also be equal to that quality level.**

Grounding for Shielded Twisted Pair

Tip 29 – **STP must be grounded because of the shield.** The ground should be connected only at one end. CAT5 STP patch panels normally provide a grounding strip or bar.

Hubs and switches do not provide grounding. If you attempt to establish a ground with an active device and it experiences an electrical disturbance, surges will occur on the cable. This will damage all equipment attached to the LAN and may create a fire hazard.

Reducing Electromagnetic Interference (EMI)

Much time and money can be saved by routing network and power cables through a single raceway. But the cables become highly susceptible to noise coupling. *Common mode voltage* signals, that is, voltage induced equally on all signal conductors by a power line, are sometimes a single-digit percentage of the power-line voltage. Because transformer-coupled systems as used on 10BASE-T and 100BASE-T reject common mode voltages, this is not a severe problem. However, these common mode problems can lead to large differential voltage and corrupt data.

Caution: *Standard Ethernet magnetics typically have 1500-V isolation ratings. IEC standard 1000-4-6 establishes a common reference for evaluating the performance of industrial-process measurement and control instrumentation when exposed to electric or electromagnetic interference. This standard requires 2000-V surge and fast transient burst tests.*

Tip 30 – **When selecting cables, it's wise to be pessimistic about their ability to reject noise from the 220 VAC and 480 VAC power lines and noisy power supplies of a factory.** CAT5 and CAT6 cables have four twisted pairs within an outer sheath to reject noise. The consistency of cable twists is vital for minimizing noise susceptibility. Good *common-mode rejection ratios* (CMRRs) of noise sources are the result of closely balanced connector and cable capacitance. Poor manufacturing tolerances, exposure to harsh chemicals, excessive physical abuse (e.g., band radiuses are too tight, there is too much stress pulling on the cable, or the cable is being run over with a forklift), and even high humidity levels can seriously degrade the capacitance balance of a twisted-pair cable.

Tip 31 – **Capacitance imbalance greater than 70 pF per 100 m can introduce harmonic distortion, resulting in bit errors.** The cable becomes much more susceptible to electromagnetic interference. Noise is induced more on one conductor than the other; it corrupts bits and causes transmission errors and retries. Ethernet cables vary by as much as 30 dB in CMRR.

Deviations from cable impedance over the length of the cable are common and negatively impact performance. This results in backward reflections and a condition known as *return loss*. Return loss is a summation of all the reflected signal energy coming backward toward the end where it originated. It is reported in decibels as a ratio of the transmitted versus reflected signal.

Return loss numbers are analogous to signal-to-noise ratio: High return loss is desirable. Low return loss means smaller negative numbers in decibels; high return loss means larger negative numbers in decibels. Selecting a cable with 5% impedance mismatch instead of 15% improves return loss by up to 10 dB.

Tip 32 – **Shielded Twisted Pair (STP) is naturally more noise-immune and is preferable to UTP in noisy situations. It should have at least 40 dB CMRR and less than 0.1-pF capacitance unbalance per foot.**

Tip 33 – **Fiber optic is certainly more expensive but it bypasses all of these electrical issues. Especially in high-speed networks, it's an attractive choice.**

Switches Are Better than Hubs

Tip 34 – **If one section of a network is exposed to excess amounts of electrical noise, it's best to isolate that section with switches.** Noise does not pass through switches, only packets headed for real destinations do. Hubs distribute messages indiscriminately and offer less protection against noise sources. If you must use a hub in a noisy environment, use one with a level of intelligence instead of a "buffer."

Better Cables Are Not Always Better

CAT6 cable can operate up to 1,000 Mbps because of its superior bandwidth. However, this can actually cause problems in high-noise environments in a 100 Mbps network because it transmits noise more easily. Ordinary CAT5 cable is better in this situation.

Don't Skimp on Cables and Connectors

The performance difference between office-grade and high-quality cables may not make the slightest difference in your house, but it could make or break an automation system that's expected to operate reliably for many years.

Tip 35 – **The cost of cable in relation to the total cost of related equipment is quite small. If you're looking for ways to save money, this is not a place to do it.** Choosing a well-designed cable will minimize your bit error rate after installation, resulting in faster throughput and fewer glitches.

Harsh Chemicals and Temperature Extremes

Tip 36 – **If your equipment is subject to washdown or exposure to corrosive chemicals, be sure to select cables with insulation rated to withstand exposure to those chemicals, such as PUR (polyurethane).** Otherwise acids, fertilizers, and petroleum can be absorbed by the cable jacket and degrade the electrical characteristics of the conductors.

Some plastics (e.g., PVC) become brittle at low temperatures, so be certain about temperature ratings. Are the cables expected to flex? Be absolutely certain you have cable that is designed for that purpose.

7.2 When You Install Cable

- If you are unable to plan the exact cable locations, add a measure of protection with armored shield or conduit.

- If physical protection or local codes necessitate using conduit, use STP wire.
- Isolate the STP shield from the conduit, since high voltages may be present on the conduit.
- Attach the STP shield to ground at only one end of the cable. Connecting at both ends creates ground loops with substantial current flow and induces noise.
- If for some reason you are required to terminate the shield at both ends, wire a metal oxide varistor (MOV), a 1 M ohm resistor and 0.01 to 0.1 μF capacitor, together in parallel. This severely limits ground current except when extreme voltages are present. See Figure 7-1.

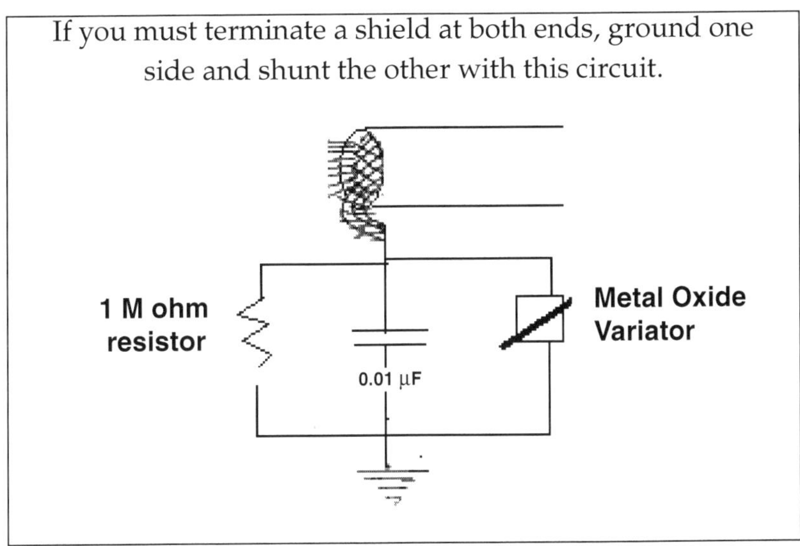

Figure 7-1. Terminating a Shield at Both Ends

- Check cables with a cable tester, not just with an ohmmeter. A tester quickly identifies continuity problems such as shorts, open wires, reversed pairs, crossed pairs, shield integrity, and miswiring of cables.

- If your cable trays are metal, they should be conductive from end to end.
- Avoid proximity to power lines and sources of electrical transients. High-voltage lines should intersect the cable at a 90° angle.
- Maintain at least a 10 cm distance from 110 VAC, 15 cm from 220 VAC, and 20 cm from 480 VAC if you use conduit. If you don't use conduit, double those distances.
- Educate unsupervised electricians about the practices described here: Purchase a copy of this book for each of them.

7.3 How to Ensure Good Fiber-Optic Connections

Communication on fiber-optic cable is greatly affected by the cleanliness of the fiber connections (see Figure 7-2), especially the cable splice and connectors. If any component is contaminated by dirt, dust, oil, etc., the transmission will be significantly degraded.

How to clean fiber-optic splices:

- Clean the fusion splice with an alcohol towel.
- Clean the connectors on S2MMs with PCB cleaner or with a cotton swab dipped in alcohol.

Fiber-Optic Distance Limits

The maximum length of a fiber-optic segment is not determined by the attenuation of the light signal but by the size of the collision domain. Exceeding allowable distances by small margins may not "crash" the network, but it will create late collisions and fragments.

These issues apply to half-duplex Ethernet with collisions. Full duplex eliminates collisions so fiber cable lengths can be greater.

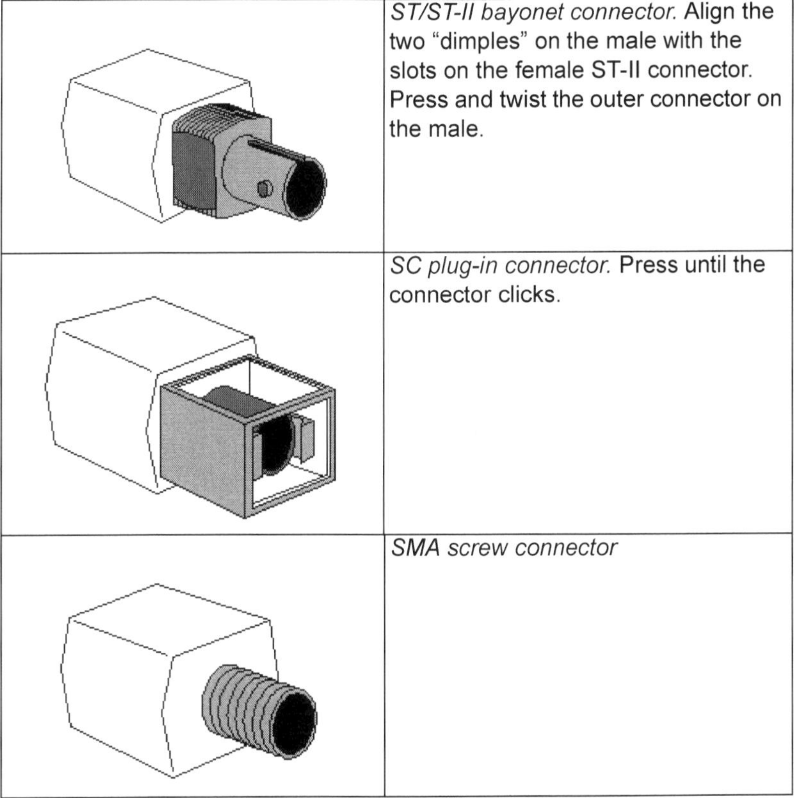

Figure 7-2. Popular Types of Fiber-Optic Connectors

With full-duplex Ethernet, the multimode limit is 2 km for 10 Mbps and 100 Mbps. Attenuation causes problems for distances greater than that. With single-mode fiber, so long as the losses in the power cable do not starve the single-mode transmitter, the installation will work.

Full-Duplex Ethernet with Single-Mode Fiber

Tip 37 – **For long distances, Ethernet needs to run in full-duplex mode. All connected segments must also be full duplex, including the switches.** Half-duplex collision domains must be connected with devices that can run the fiber link in full-duplex mode (e.g., a bridge, or using one port of a switch in full-duplex mode).

8.0

Ethernet Industrial Protocols, Fieldbuses, and Legacy Networks

I can't tell you how many times over the years I've had someone ask "what is the best Ethernet protocol?" That's like asking who has the best pancakes, what is the most exciting sporting event, and what is the best Christmas movie of all time? (FYI, the answers are: 1. Pancake Place in Green Bay, Wisconsin, 2. NCAA Basketball Tournament, and 3. *It's a Wonderful Life*.)

The truth, of course, on the question of factory floor protocols is simply "it depends." Do you want to move I/O data or information? How fast do you need to move the data? How much data do you need to move? How many devices have the data now? There is no end to the questions you should ask and answer.

In practice, it is not that complicated. In the great majority of cases, it just comes down to the brand of programmable logic controller (PLC) used in your building, or used by the majority of your customers, that drives the Ethernet protocol. If you are using Siemens PLCs, the "best" Ethernet protocol is PROFI-

NET IO. If you're using a Rockwell ControlLogix or Compact-Logix, it is EtherNet/IP. If you want to keep everybody happy, or at least not too unhappy, use Modbus TCP.

From a purely technical standpoint, Ethernet and any of these Ethernet protocols are not necessarily more ideal for automation applications than any other network. Even if there was one that was superior to the others, the nature of capital equipment is still such that no one is going to rip out existing equipment and wiring just because something better exists.

But ignoring the PLC brand issue, there is a series of questions that you could ask to determine if you should use an Ethernet protocol or another type of industrial communication system (CAN, Modbus Serial, PROFIBUS DP, etc.). These questions include:

- What is the distance requirement?
- What kind of physical cabling arrangement makes sense for this application? All the Ethernet formats except 10BASE2 and 10BASE5 use a star topology. This is fine for applications where devices are clustered together in groups but for others, such as a long conveyor with many nodes spaced 20 m apart, it is quite inconvenient. For the conveyor, a trunk/drop topology (such as that used by DeviceNet™ and CANopen) is much better.
- What is the actual speed (response time) requirement for the most time-critical devices? Do all of the devices require that level of speed or should some devices have a higher priority than others?
- Does your application require that you prioritize messages?

- Do the devices you want to use support the same network standard? Are there open versus closed architecture considerations?
- If you are developing a network-capable product, what is the hardware bill of materials and the cost of software development for that network?
  - How much electrical noise is present in the application and how susceptible is the cabling?
  - What is the maximum required packet size for the data you are sending? If the data can be fragmented over several packets, how fast does a completed message have to arrive?
  - What types of device relationships are desired (master/slave, peer-to-peer, broadcast)?
  - Does the network need to distribute electrical power? If yes, how much current?
  - What kind of fault tolerance must be built into the network architecture?
  - What is the total estimated installed cost?

With the answers to these questions you can make a reasonably good choice on an industrial networking protocol.

8.1 The Two Most Important Points to Understand

This chapter describes the leading industrial automation protocols. To understand those protocols, you need to understand two key points that apply to all industrial Ethernet protocols. The first key point is that the most important differentiator from one Ethernet protocol to another is the data representation. The data representation—how data is organized in devices and implemented in the address space of the protocol—is what makes the protocol unique. It is the key to a genuine understanding of any of these technologies. Everything else that describes the protocol—how data is transported, how connections are made, and what services exist to provide one device (usually the client) with access to the data in the address space of another device (usually the server)—is actually similar from protocol to protocol. For example, both EtherNet/IP and PROFINET IO use synchronous messaging and both EtherNet/IP and Modbus TCP make more extensive use of TCP. There are a lot of commonalities but the way data is accessed in the address space of a device is the key differentiator.

The second key point is that industrial Ethernet protocols are simply a way of defining messages that pass through the "pipe" known as the *TCP/IP stack*. You can think of a TCP/IP connection between two devices as a phone connection. The connection can exist even if no one is talking (sending data) over the connection. That is why these Ethernet protocols are known as *application layer protocols,* they are the "applications" that uses the TCP/IP stack.

Each of the industrial protocols make use of the TCP/IP stack and TCP/IP connections in different ways, but they all send messages through the pipe in one way or another. All protocols use the IP layer. Some solely use TCP (Modbus TCP), some use both TCP and UDP (EtherNet/IP), and one uses TCP but also has another channel that bypasses the TCP/IP layer.

When reading the following sections on these Ethernet application layer protocols, make special note on how the data is organized, the object model of the protocol, and how each protocol makes use of the services of the TCP/IP stack.

8.2 Modbus and Modbus TCP

Saying that you want to discuss Modbus can sometimes bring to mind chats about buggy whips, the rotary telephone, or that new innovation, the color television. What is there to say? What hasn't been said about Modbus over the last 40 years?

Modbus is hardly a new technology. Historians can disagree about its actual birth, but it is certainly a product born in the 1970s. Success is such a trite word for how well it has done over those 40 years. Modbus has found its way into hundreds of thousands—if not millions—of devices. You can find it in everything from valve controllers, to motor drives, to human-machine interfaces (HMIs), to water filtration systems. It would be difficult indeed to name a product category in industrial or building automation that does not use Modbus.

Yet even in the automation world, Modbus isn't just old technology. ***It is ancient technology.*** Modbus is like that lovable old uncle that comes over every Thanksgiving. He's retired now, he putters around his garden, he's no longer the handsome debonair man of 40 years ago, but he's there when we need him and that is why we love him.

Prior to Modbus, all we had was electrical signaling. Modbus changed that. In fact, Modbus changed everything. Modbus introduced the concept of data on the factory floor. Modbus made it possible to connect an entire group of devices using only two wires on the controller. That alone saved a massive investment in wire, labor, and installation time. Instead of miles and miles of wire connecting hundreds of devices, a simple two-wire pair could be used to daisy-chain devices together. It was revolutionary for its time.

It wasn't just that Modbus was the first serial protocol. Modbus was the right technology at the right time. Remember that the first microprocessor wasn't invented until shortly before the birth of Modbus. Do you remember what those microprocessors were like? Simple 8-bit processors with severely limited code space and memory.

Modbus is the most pervasive communications protocol in industrial and building automation and the most commonly available means of connecting automated electronic devices. Why did that happen? Why did Modbus have such an impact on the industrial automation industry that it has survived for over 40 years and is to this day one of the leading industrial networks of the twenty-first century? There are three primary keys to its success:

- **Modbus Is an Open Standard** – Modicon, the inventor of Modbus, did not keep the standard proprietary. They released it as a nonproprietary standard and welcomed developers, even competitors, to implement it. They rightly assumed that it would be best for everyone, including them, if Modbus became successful in the marketplace. Because of this thinking, Modbus became the first widely accepted fieldbus standard. In a short time, hundreds of vendors implemented the Modbus messaging system in their devices and Modbus became the de facto standard for industrial communication networks.
- **Modbus Uses Standard Transports** – The transport layer for Modbus remote terminal unit (RTU) commands is simply RS-485, a differential communication standard that supports up to 32 nodes in a multi-dropped bus configuration. The RS-485 standard provided noise immunity that was superior to that in the RS-232 electrical standard.

The transport layer for Modbus TCP commands is TCP and only TCP.

- **Modbus is a Simple Protocol** – Modbus is quite easy to understand (see Figure 8-1). Its primary purpose is to simply move data between an RTU master device (a *client* in Modbus TCP) and one or more RTU slave devices (*servers* in the Modbus TCP world). There are only two kinds of data to move, register data, and coil data. Registers are 16-bit unsigned integers. Coils are single bits.

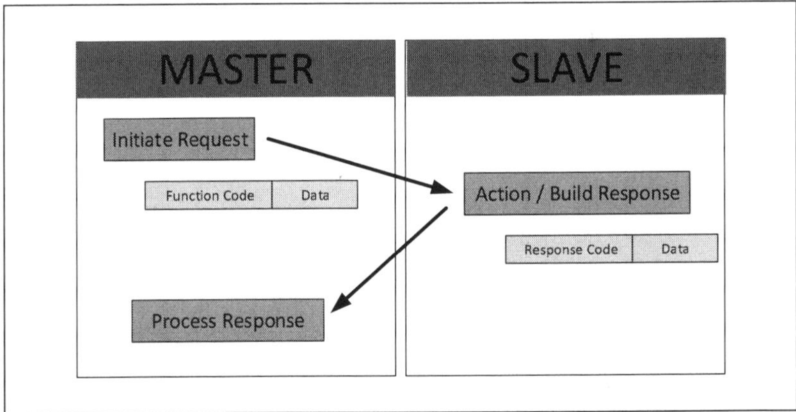

Figure 8-1. RTU Master/RTU Slave Modbus Architecture

Modbus uses a straightforward request/response command structure. A Modbus master requests or sends data to a slave and the slave responds. There are simple commands to read a register, read a coil, write a register, and write a coil

Modbus TCP and Modbus serial use ***exactly*** the same byte sequences to implement the command/response. The typical components of a Modbus message are presented in Table 8-1.

Table 8-1. Modbus Message Components

Function Code (FC)	The function code identifies the request to the Modbus slave. There are a large number of possible message requests, but about eight that are commonly used. These are the function codes that are detailed in this chapter.
Starting Address	The starting address is the index into the data area in the Modbus device. If the function code targets coils, this field specifies the index into the coils (bits) of the coil address space. If the function applies to registers, this field specifies the index into the registers for that part of the address space. Note: Modbus address spaces are one-based—the first register or coil is one. The Modbus protocol is zero-based. The first register or coil is zero. The address on the wire is always one less than the address in the Modbus data request.
Bit Length	The number of bits to read or write.
Word Count	The number of registers to read or write.
Byte Count	The number of data bytes included in the message request or response.
Response Code	This byte indicates the successful completion of the message request. It is identical to the original message request.
Exception Response (FC)	An exception response is indicated by combining the response code of the original Modbus function request with 80 hexadecimal. For example, a Modbus exception response to function code 3 is 83 hexadecimal. A single data byte value with the Modbus error code always follows the exception response byte.

For Modbus TCP, this set of message components is inserted as the data bytes of a standard TCP message as shown in Figure 8-2.

Modbus messages can be encoded, meaning turned into a series of bits, in one of two ways: Modbus ASCII or Modbus RTU. Modbus ASCII is a relic of the days of teletypes; every

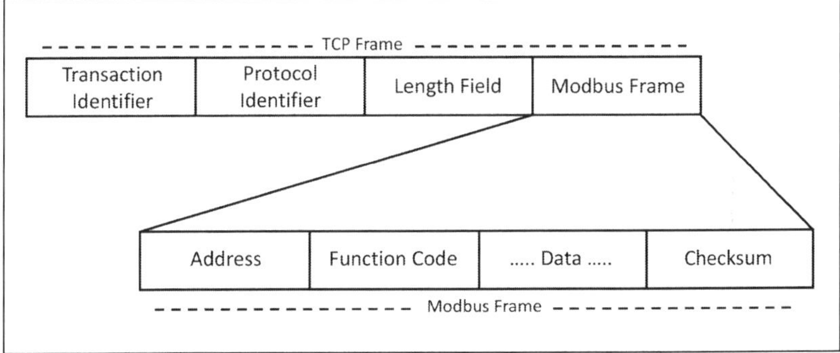

Figure 8-2. Format of a Modbus/TCP Frame (Courtesy Schneider Automation, www.modbus.org)

byte is transmitted as two ASCII characters. Few devices today use Modbus ASCII. Modbus RTU, on the other hand, is very popular and almost all Modbus serial and Modbus TCP devices use it. Modbus RTU encodes each byte of a message as a binary data value, which greatly enhances performance over Modbus ASCII. Modbus TCP devices always use Modbus TCP.

Modbus on a *serial* network is not fast—response times of a fraction of a second are not at all uncommon. Which, of course, is why Ethernet is an attractive alternative to serial—10, 100, or even 1,000 Mb performance is possible. Also, the simple Modbus protocol does not support complex objects and sophisticated device profiles. The master/slave orientation does not prevent peer-to-peer communication, but it requires separate "sessions" to be opened up between devices.

The power of Modbus has always been its simplicity. Modbus fit well in the era of limited RAM and FLASH. It required little code space (FLASH), often as little as 1K. Memory (RAM) varies with the size of the Modbus data space that you needed to represent the device's data. Simple automation devices with little bits of data—imagine a photo eye—could be implemented with hardly any RAM space. These devices could now, for the first time, send their data to a control system as part of a daisy-chained 485 network, avoiding hardwired point-to-point communications.

The simplicity of Modbus has been both a blessing and a curse over the years. The simplicity has led to an incredible amount of activity and propagation of Modbus into many different industries around the world. There is probably no product category in the last 40 years that has not had an offering without Modbus.

The simplicity of Modbus has also led to many companies expanding the message structure, data representation, and transports. Some vendors have imposed any number of advanced structures and data types on the basic Modbus address structure. Others have used Modbus in other ways that go beyond the basic specification. These implementation extensions are not expressly prohibited by the specification, but they do not always make Modbus easily portable to many different applications.

8.3 EtherNet/IP

EtherNet/IP is an industrial Ethernet application layer protocol used by Rockwell Automation (Allen-Bradley) programmable controllers. EtherNet/IP uses standard Ethernet to organize the task of configuring, accessing, and controlling industrial automation devices. What does that mean? It means that EtherNet/IP is the highly structured protocol that uses Ethernet to move inputs from industrial end devices into an Allen-Bradly programmable controller and moves outputs generated by the control logic of an Allen-Bradley programmable controller to devices that map those outputs to real world physical outputs.

EtherNet/IP is based on the Control and Information Protocol (CIP) used in DeviceNet, CompoNet™, and ControlNet™. CIP provides a common, standardized mechanism for representing data, sending messages, and defining common device types for the component technologies that use the CIP core protocol. CIP is a media-independent protocol, which means that CIP messages can be sent over any communication media including CAN, Ethernet, and even something like FireWire. Sending CIP messages over CAN forms the basis for the DeviceNet protocol. Sending CIP messages over the ControlNet communication bus is the basis for ControlNet. Sending CIP messages over Ethernet TCP and UDP is the basis for EtherNet/IP. CIP provides the core technology used in each of these application layer protocols.

CIP defines two kinds of messages: explicit and implicit. Explicit messages are asynchronous, request/response type messages. A sender builds a request and sends it to a receiver. The receiver receives the request, opens it, decodes it, and sends a response. It is the traditional mechanism for communication between two devices. Implicit messages are synchronous messages that are continuously passed back and forth between the sender and receiver. Unlike explicit messages, the contents of implicit messages are simply raw data. Both the sender and the receiver have to have prior knowledge of how to construct and decode that raw data. How the sender and receiver map that implicit data is described later in this section.

The CIP protocols—EtherNet/IP, ControlNet, and DeviceNet—all define a type of controller device and some type of end device. Unlike other CIP protocols that use the terms master and slaves, the "master" device in EtherNet/IP is labeled a *scanner* and end devices, instead of slaves, are labeled as *adapters*. Scanners, typically programmable controllers, open connections with adapters, configure the timing of the asynchronous implicit messaging with the adapters, and send explicit messages to adapters when needed. Adapter devices are typically industrial I/O devices, such as valves, I/O blocks, drives, scales, meters, and other end devices you might find in an automation system. An adapter has one job: send the scanner an implicit message with the status of its real world inputs and set its real world outputs as directed in the implicit output message received from the scanner.

EtherNet/IP uses TCP/IP for explicit messaging. Explicit messages (messages that are sent asynchronously) are sent over TCP. Examples of explicit messages include: changing the ramp time on a drive, setting a tare weight on a scale, and reading a barcode using TCP as the initiator of the message. By using TCP, the initiator automatically gets delivery acknowledgement for these important messages. On the other hand, implicit messages (messages that are delivered synchronously) are sent over UDP because they do not require delivery notification. By definition, a lost synchronous message is going to be replaced quickly by the next message in the sequence.

One of the most important features of EtherNet/IP (and CIP in general) is how it models device data in adapter devices. EtherNet/IP devices—in fact all CIP devices—are modeled as a collection of objects, each containing related data (a model of an EtherNet/IP device is illustrated in Figure 8-3). Objects are composed of data values or *attributes* in CIP terminology. Attributes values can be assigned a type with any one of a large number of EtherNet/IP types to model the specific data in the device.

The object nature of EtherNet/IP does not imply that the device implements the object structure internally, only that the device looks to the EtherNet/IP network as a collection of objects each with one or more attributes. These attributes form the available data that an EtherNet/IP device exposes to the outside world. Scanners can access these attributes using explicit or implicit messaging.

There are two kinds of objects in every EtherNet/IP device: required objects and application objects (see Figure 8-3). Required objects must be present in every EtherNet/IP device while application objects are particular to the function of the end device. For example, every EtherNet/IP device must have an identity object, an Ethernet object, a TCP object, a router object, and a connection object. Each object provides attributes that describe the specific functionality of the device. The identity object, for example, presents identity information to the network by making available attributes like the vendor ID, the product code, the software revision, and other information that specifically identifies that device and its application. The TCP object provides information on the TCP connection like the TCP/IP address of the device. The connection object provides information on the current connections to a controller.

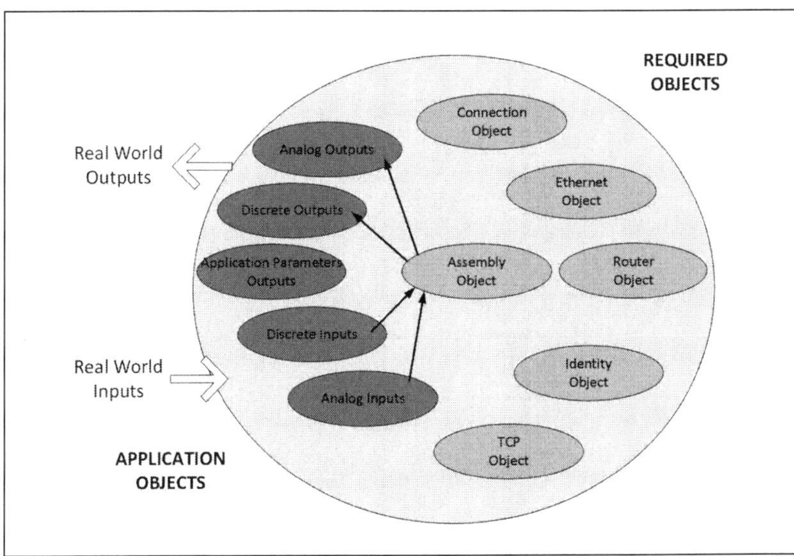

Figure 8-3. EtherNet/IP Object Representation

The object number and attribute numbers of required objects are predefined and identical in every EtherNet/IP device. Object 1 is always the identity object, Object 2 is defined to be the router object and so on. Attributes for the required objects are also predefined. Attribute 1 of the identity object is always the vendor ID, Attribute 2 is the device type and so on. By predefining the object numbers and attribute numbers for all the required objects, a controller or a PC tool always knows exactly how to get specific information about an EtherNet/IP device or, in actuality, any CIP device.

Application objects are the set of objects that model the I/O data of the adapter device. The set of application objects can be simple or complex. The object model for a simple device like an 8-channel valve might simply be one application object with one 8-bit attribute containing the current status of the valve states and one 8-bit attribute containing the commanded state of each valve as currently specified by the scanner. For a more sophisticated device, like a motor drive, there might be tens or even hundreds of objects to provide access to all the functionality of that device. The complexity of the object model in an EtherNet/IP device is directly related to the complexity of the data being exposed to the network through the object interface.

These application layer objects are predefined for a large number of common device types. All CIP devices with the same device type (drive system, motion control, valve transducer, etc.) must contain the identical series of application objects. The series of application objects for a particular device type is known as the *device profile*. A large number of profiles for many device types have been defined. Supporting a device profile allows a user to easily understand and switch from a vendor of one device type to another vendor with that same device type.

A device vendor can also group application layer objects into assembly objects (Figure 8-4). These super objects contain attributes of one or more application layer objects. Assembly objects form a convenient package for transporting implicit messages between devices. For example, a vendor of a temperature controller with multiple temperature loops may define an assembly for each temperature loop and an assembly with the data for all temperature loops. The user can then pick the assembly that is most suited to the application.

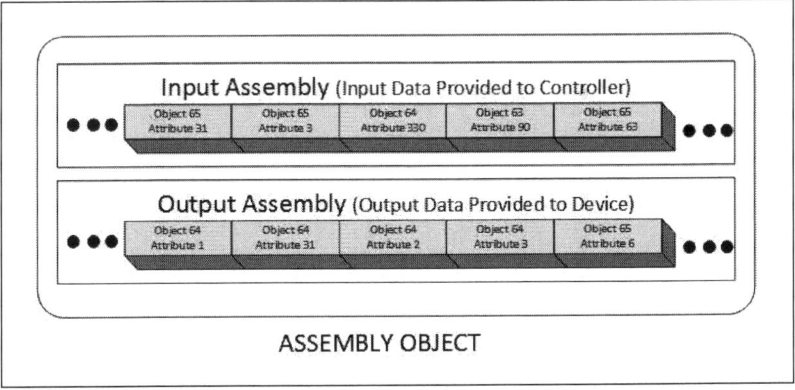

Figure 8-4. Assembly Object Structure

Assemblies are what is transferred in the implicit message. The input assembly is the set of attributes that are delivered to the scanner each time the implicit message is triggered. The output assembly is the set of attributes that are received from the scanner on each implicit output message. The contents of these assemblies are specified in an EDS or Electronic Data Sheet. The EDS can be used by the engineer configuring the controller to assemble the data to deliver in the output assembly and decode the data received in the input assembly. Another mechanism to decode the implicit message assemblies is the Add-On Profile (AOP). The AOP is a way of electronically configuring a

Rockwell Controller to know how to encode and decode messages for a specific device.

The Open Device Vendor Association (ODVA), headquartered in Ann Arbor, Michigan, is the vendor trade association that manages all CIP technologies, including EtherNet/IP. ODVA members, some of the world's leading automation companies, work to advance the development of CIP technologies and promote interoperability among vendor devices. One of the most important jobs of the association is conformance testing. Vendors manufacturing EtherNet/IP devices must submit each new device for conformance testing at the ODVA test lab. In the test lab, each device is exercised independently and in a rack containing a multitude of other vendor devices. The long sequence of tests verifies not only that the device adheres to the ODVA specification, but that it interoperates with other EtherNet/IP devices from other manufacturers. Once certified, the manufacturer can exhibit the conformance logo (Figure 8-5) indicating to users that the device is certified.

Figure 8-5. EtherNet/IP Conformance Tested Logo

EtherNet/IP is a widely implemented protocol with numerous advantages. First, Ethernet/IP uses the tools and technologies of traditional Ethernet. It uses all the transport and control protocols of standard Ethernet including the Transport Control

Protocol (TCP), the User Datagram Protocol (UDP), and the media access and signaling technologies found in off-the-shelf Ethernet. Building on these standard IP technologies means that EtherNet/IP works transparently with all standard off-the-shelf Ethernet devices (switches, routers, diagnostic tools, etc.) found in today's marketplace. It also means that EtherNet/IP is easily supported on standard PCs and all their derivatives. But even more importantly, because EtherNet/IP is based on a standard technology platform, EtherNet/IP will move forward as the base technologies evolve in the future. Secondly, as discussed above, Ethernet/IP is a certifiable standard. Devices are tested in a lab to verify that they meet the EtherNet/IP standard, which ensures interoperability between devices from multiple vendors and the consistency and quality of field devices. This ensures the consistency and quality of field devices. Third, EtherNet/IP is built on the widely accepted CIP protocol layer. Finally, and most importantly, EtherNet/IP is the industrial application layer protocol that is used by Rockwell Logix programmable controllers to communicate with Ethernet-enabled field devices. With Rockwell programmable controllers having a significant share of the programmable controller market in North America, EtherNet/IP will continue to dominate the industrial application landscape for years to come.

8.4 PROFINET

This is the PROFIBUS Trade Organization's answer to the need for interoperability between automation devices and subsystems that are linked together via Ethernet.

To understand what PROFINET is, you must understand what it is not. PROFINET is not the PROFIBUS protocol on Ethernet in the same way that Modbus/TCP is the old familiar Modbus on Ethernet. PROFINET is not really a "fieldbus" as the term is normally understood, either.

PROFINET is not even Ethernet-specific; it links via TCP/IP and occupies layers 3 and above in the ISO/OSI model. Other physical layers, such as modems, WANs, VPNs, or the Internet may be employed so long as a PROFINET device is linked to the network via TCP/IP. An analogy to the office environment may help you understand what it is intended to do.

The PC in your office at work is networked with a dozen other PCs and a file server. Your office LAN (Ethernet 100BASE-T) is also linked to a T1 Internet line. You open Microsoft Word and create a complex document. You write some text and create some tables. Your coworker Jeff has a PowerPoint presentation on his PC; you open it via the network and copy and paste two graphics images into your Word document—the images are transferred intact as objects. Your other coworker Leslie has an Excel spreadsheet that you also open remotely and embed in your document—it is as simple as cutting and pasting. In this case, the Excel data is not static, it is live. Leslie updates this spreadsheet every Tuesday, and every time you open your document it will retrieve the latest data from her document on her PC. Finally, you access the Internet, copy and paste text and graphics from one website to your document, and then insert hyperlinks to other websites.

Behind this transparency among applications is a very complex object model created by Microsoft. Savvy PC users are accustomed to this level of sophistication and its benefits. This expectation naturally extends to the integration of business applications throughout an entire company and, of course, to devices in an automation system. This is the expectation that

PROFINET was engineered to satisfy. The OLE[1] for Process Control (OPC) software standard (www.opcfoundation.org) was developed to create transparency between hardware devices (e.g., network and I/O cards) and software applications (operator interface and programming tools). PROFINET uses components of OPC (COM and DCOM) and extends this transparency to all devices on a TCP/IP network, further defining object models for many kinds of device and programming parameters.

Rather than being specific to only one manufacturer's hardware or software (as is often the case with Microsoft), PROFINET is an industry standard available to all PROFIBUS members. PROFINET is an open communications and multivendor engineering model. This means that a preconfigured, preprogrammed, and pretested machine such as a transport conveyor can be set up using the vendor-specific electrical devices and applications as it has been in the past.

With PROFINET, the entire vendor-specific module (machine, electrical, and software) is represented as a vendor-independent PROFINET component. This PROFINET component is described within a standardized XML file that can be loaded into any PROFINET engineering tool, and interconnections between the PROFINET objects can be established by connecting lines from object interface to object interface.

In regards to communication and physical topology, established protocols such as TCP/IP, RPC, and DCOM are used. Data access to the PROFINET objects is standardized via OPC. As for physical device connections, not only can devices be connected via an integrated PROFINET interface, but existing

1. OLE stands for Object Linking and Embedding, a standard developed by Microsoft.

intelligent devices that are currently used with fieldbus networks, such as PROFIBUS, can be connected to Ethernet through a gateway device called a PROFINET *proxy server*.

Every PROFINET object is described by an XML file that defines these parameters so that every defined data type in the system is accessible by name throughout the PROFINET network. Integrators do not have to link devices at the bit level. It is expected that PROFINET proxies for each different fieldbus system (PROFIBUS, DeviceNet, Modbus, ControlNet, and others) will be developed over time, extending the transparency of large systems.

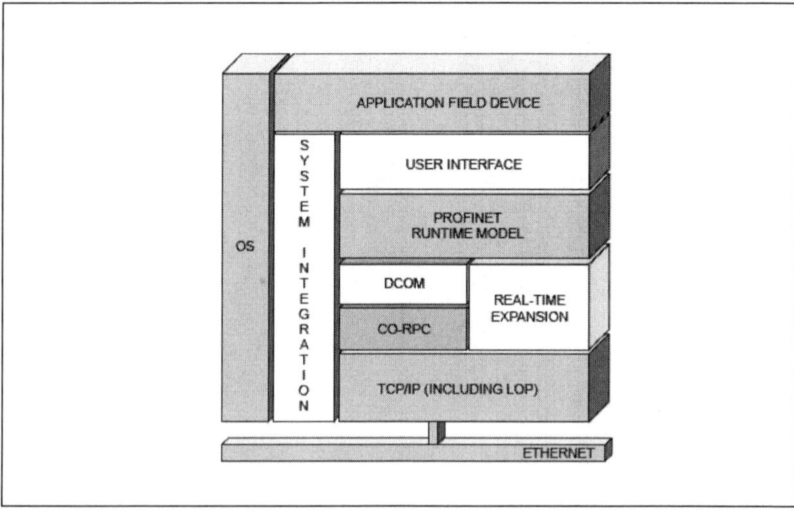

Figure 8-6. The Communication Layers of PROFINET

To delve into the internals of PROFINET as described in Figure 8-6 is beyond the scope of this book. However, more information is available at www.PROFIBUS.com, and PROFIBUS organization members can download the specification and source code at the site.

8.5 FOUNDATION Fieldbus High-Speed Ethernet

This protocol uses the FOUNDATION Fieldbus H1 process control protocol on TCP/IP. FOUNDATION Fieldbus H1 is a sophisticated, object-oriented protocol that operates at 31.25 Kbps on standard 4–20 mA circuits. It uses multiple messaging formats and allows a controller to recognize a rich set of configuration and parameter information (device description) from devices that have been plugged into the bus. FOUNDATION Fieldbus even allows a device to transmit parameters relating to the estimated reliability of a particular piece of data. FOUNDATION Fieldbus uses a scheduler to guarantee the delivery of messages, so issues of determinism and repeatability are solidly addressed. Each segment of the network contains one scheduler.

FOUNDATION Fieldbus high speed Ethernet (HSE) is the same as the H1 protocol, but instead of 31.25 Kbps, it runs on TCP/IP at 100 Mbps. It provides the same services and transparency of network objects but operates at a higher level.

FOUNDATION Fieldbus is specifically focused on the process control industry and will likely be the dominant Ethernet I/O standard there.

Installations in this segment of the world typically have the following characteristics:

- Very large campuses (e.g., chemical refineries) with many nodes
- Data does not have to move quickly, but there is a lot of it to move (large packets)
- Large quantities of analog data
- Hazardous area classifications such as Class I, Division 2

9.0

Basic Precautions for Network Security

Disclaimer

The topic of information security is complex and expansive enough to warrant an entire book, let alone a small chapter. A quick search on "network security" at www.amazon.com turned up 37,534 books. That's more than a good weekend of reading for anyone. The scope of this chapter is purposely constrained to fit within the pages allocated. It focuses solely on an overview of industrial security, something akin to learning to fly an airplane by looking through the window in the departure lounge.

All disciplines have a language of their own. Network security is no exception. To discuss network security—the threats, attack profiles, and security features to counter those threats—it is helpful to understand some basic terms:

- **Public Key** – A series of bytes which form a key that the owner makes available to anyone who requests it.

- **Private Key** – A series of bytes which form a key that is kept private by the owner and never released to anyone else.
- **Digital Certificates** – A sequence of data bytes that functions like a driver's license. The digital certificate verifies that you are who you say you are. There are many components to a digital certificate, including the name of the algorithm and the organization that created it, the owner's public key, and the dates it is valid. *X.509* (also X*509 certificates)* refers to the most popular certificate standard. You will also encounter the term *Distinguished Encoding Rules (DER) certificates,* which refers to the method for encoding certificates as a binary series of bytes.
- **Certification Authority (CA)** – An organization that creates and distributes digital certificates. The CA creates the public and private keys that are associated with the certificate owner. The CA often encrypts a portion of the certificate with its private key (i.e., signs it) to assure everyone that the CA did create the certificate. Of course, sending a certificate to a receiver is only effective if the receiver knows the CA is an honest and reputable certificate provider.
- **Digital Signature** – Also called *signing,* a digital signature is a small series of bytes that results from processing a larger series of bytes through an algorithm. The resulting smaller series of bytes is encrypted with the owner's private key. Using the owner's public key and validating the result with the sender's same algorithm, the receiver can decode the encrypted result and verify that the true owner "signed" the document. Signing a document or message guarantees the integrity

of the portion of the message signed with the owner's private key.

- **Public Key Encryption** – An encryption process in which private and public keys are exchanged to sign and encrypt/decrypt messages.
- **Public Key Infrastructure (PKI)** –The set of hardware, software, and policies needed to manage certificates, keys, access lists, and keys used in public key encryption.
- **Authentication** – Verifying who the sender is. Usually that means validating the sender's X509 certificate and verifying that the certificate is currently valid and that it is signed by a reputable and trusted CA.
- **Authorization** – The process of validating access to a resource. Once a sender is authenticated, the sender must be granted access to resources by the receiver. Authorization can be accomplished using a list of trusted names, a user name and password, or any other reliable mechanism.
- **RSA** – A popular public key cryptography algorithm. RSA refers to the initials of the three designers of the algorithm: Rivest, Shamir, and Adleman. RSA, with its variously sized key lengths, is used in OPC UA and other popular and secure protocols.
- **SHA** – A series of public key cryptography algorithms published by the National Institute of Standards and Technology (NIST). SHA algorithms are also used in OPC UA to sign and encrypt messages.
- **Auditing** – The recording of all actions, activities, users, resources, and more in a system. In the OPC UA architecture standard (see Chapter 14), auditing is

incorporated as a normative part of the specification as a mechanism for system administrators to use to identify vulnerabilities and diagnose security breaches.

- **Symmetric Security** – Security in which both the receiver and the sender hold the key to decrypt messages. One encrypts with the symmetric key, the other decrypts with it. It is called *symmetric* since both hold the same key.
- **Asymmetric Security** – Security in which both the sender and the receiver have a private key that they keep secret and a public key that they share with anyone. Messages to the key holder are encrypted with the public key and decoded by the key holder with the private key. Message segments that require verification of origin are signed with the private key and validated using the public key.

The threats to industrial networks, sometimes called *attack vectors,* are many, some of the more common ones are:

- **Message Flooding** – An attack in which a significant volume of message requests, both well-formed and malformed, are delivered to industrial Ethernet devices, overwhelming their capabilities to respond to valid messages.
- **Eavesdropping** – The unauthorized disclosure of any information about your processes, operation, or technology. Often eavesdropping provides information that leads to a more serious follow-on attack.
- **Message Spoofing** – The sending of messages that appear to be from a legitimate source that direct an

Ethernet device to perform unauthorized operations or actions.

- **Message Alteration** – The modification of messages in-transit to cause unauthorized operations or actions.
- **Message Replay** – The resending of one or more previous messages to cause a previously authorized action to occur at another point in time where it is not authorized.
- **Malformed Messages** – The sending of a series of messages that are incomplete, structured improperly, or contain excess data bytes to confuse a device into acting improperly.
- **Session Hijacking** – An attacker analyzes a running session between two applications and injects unauthorized messages into the message stream or completely hijacks the application session.

There are sophisticated mechanisms for meeting these threats, many too complex to detail here, but there are also simple guidelines you can follow to implement a modicum of network security. These guidelines are:

1. Never mix your office LAN with your industrial-control LAN. Always separate the factory network from the enterprise LAN by a firewall or, at minimum, a bridge or router. A control network and a business LAN have two entirely different purposes and their interaction should be closely controlled.
2. Focus as much on your people procedures as you do on the technology. Most security intrusions are the result of well-intentioned employees, not outsiders.
3. Do not connect consumer plug-and-play devices to a

factory LAN. A printer, for example, might flood the network with traffic with a "broadcast storm" as it tries to self-configure or advertise its presence to all nodes on the network.

4. Faulty devices, such as defective NIC cards, can emit massive volumes of bad packets (i.e., runts, which are abnormally short Ethernet frames) into a network. Use switches instead of hubs to limit the effects of such problems. Diagnostic tools can be used to locate the source of bad traffic.

5. In many industrial systems, TCP/IP address are not allocated automatically. Usually system designers fix the addresses of every device on the network. Humans being human, sometimes two devices will have the same TCP/IP address. This can easily occur when devices are replaced, and it is a perplexing problem to trace. Implement good procedures and documentation to limit this problem.

6. Change your passwords on a regular basis. Passwords that stay the same for years are often easy to guess.

7. Install routing switches or level 3 switches that can logically divide networks on the basis of IP address, IP subnet, protocol, port number, or application, completely blocking traffic that does not fit a precise profile. This offers substantial protection against broadcast storms and faulty packets while allowing specific data to freely pass between the business LAN and the factory network.

8. It's unwise to assume that your industrial Ethernet products have any security features at all. At a minimum, you should use inspection-type firewalls (such as packet filters) to control access. These firewalls

control access using a combination of the IP source address, destination address, and port number. This is by no means completely hacker proof, but it should keep the well-meaning employees out.

Those are the simple guidelines. To develop a more secure network infrastructure, there are concepts that you should consider. These include:

- **Auditability** – An important aspect of network security is the ability to trace network activity to detect intrusions and evaluate the effectiveness of the network security. Make sure that the network security technology you implement has an audit/trace feature.
- **Authentication** – Identify technologies that require entities on your network to have a mechanism for proving their legitimacy. A control valve, for example, should have a mechanism to be certain that it is connected to the actual operator console and not a rogue device.
- **Authorization** – Beyond knowing that a device is connected to authenticated devices (devices who are who they say they are), look for systems that contain a mechanism for the device to know which devices can access its resources and specific levels of services. A valve can legitimately be connected to many devices that want to know if it is open or closed. It must have a mechanism to determine if a device is authorized to access its resources (i.e., initiating an open or close operation). Authorization can be coarse or fine grained. Some systems may authorize any authenticated device to perform any operation while others may provide sophisticated authorizations where only specific devices can perform specific operations at specific times.

- **Confidentiality** – Look for systems that ensure that private information on your industrial network can be protected from unauthorized access. You will not want to encrypt every message, but critical information that you would not want released to competitors and others should be encrypted.
- **Integrity** – Integrity is the term for knowing that a message has not been altered during transmission. At a minimum, use simple checksums. Checksums provide a small modicum of security. For additional security, look for technologies that offer digital signing. In digital signing, a message is hashed and the hash key is encrypted. The receiver recomputes the hash key to validate the integrity of the message.

10.0 Power over Ethernet (PoE)

Power over Ethernet technology allows Ethernet devices to receive power and data over their existing LAN cabling without modifications to any of the existing Ethernet infrastructure. PoE radically reduces installation costs by eliminating the conduit, power wires, and installation labor required to install an Ethernet device.

10.1 What Is PoE?

Everyone who has ever used a standard phone line is familiar with a network-powered device. Simply plug the phone into the jack and you have a network connection. Power over Ethernet, IEEE standard 802.3af, performs the same function for Ethernet devices. In fact, PoE will really enable the widespread use of IP telephones.

PoE brings the ability to connect and power devices like barcode readers, RFID Systems, wireless access points, and web-based security cameras. PoE enables a whole new generation of networked devices. Because there is no need for the device to be anywhere near a wall socket, we can expect a plethora of innovative applications, from vending and gaming machines to building access systems and retail point-of-sale systems.

There are two types of PoE devices: Power Sourcing Equipment (PSE) and Powered Device (PD). PSE devices supply network power to PD devices. Before supplying network power a PSE device must first determine if the end device is PoE-enabled. Power is never delivered to a non-PoE-enabled device.

10.2 What Pins Are Used on the CAT5 Cable?

A standard CAT5 Ethernet cable has four twisted-wire pairs. Only two of these pairs are required for 10BASE-T and 100BASE-TX operation.

The PoE specification provides two ways for a PSE device to deliver power:

- **Option 1** – Power is supplied on the two spare pairs. Pins 4 and 5 are tied together to form the positive supply while pins 7 and 8 are tied together to form the negative supply.
- **Option 2** – Power is supplied on the data pairs, the same lines carrying your Ethernet messages. Since Ethernet pairs are transformer coupled at each end, it is possible to apply DC power to the center tap of the isolation transformer without affecting the data transfer. In this mode of operation, the pair on pins 3 and 6 and the pair on pins 1 and 2 can be of either polarity.

All PSE devices are required to support both power options. A PD device can support either option.

10.3 How Much Current Is Supplied?

A lot of work was done to determine how much voltage could be applied to an Ethernet device. There were several constraints on the system design:

1. PoE devices must not present a safety hazard. The voltage must be low enough to preclude an electrical shock.
2. No changes to existing cable.
3. No interference to the network data transfer.
4. No damage to non-PoE-enabled devices.
5. No requirement for extensive testing or certification of PoE devices.

After much analysis the IEEE standard committee picked a 48-VDC system that could deliver up to 12 W of power to a PoE-enabled device. At this voltage, common electrical standards do not require certification and testing, CAT5 cable is an acceptable delivery medium, and there is enough power to provide a decent functionality in the end device. At this voltage, a PD device can count on 350 mA at 37 V (12.95 W) after cable and other power losses are deducted. This number is a function of the limitations of CAT5 cable and the lack of a requirement for stringent device testing below 57 V.

To prevent power delivery to non-PoE-enabled devices, the standard includes a mechanism that a PSE can use to interrogate a device to determine if it is PoE enabled. This "discovery" mode allows a PoE to maintain compatibility with existing, non-PoE-enabled, equipment.

In "discovery mode" a PSE applies a small current-limited voltage to an end device. The PSE is looking for the presence of a 25 K ohm resistor in the remote device. If it finds the resistor, power is supplied. If not, the PSE marks the port as non-PoE-enabled and power is not applied to that device. The PSE continuously repeats the test in case a new device is connected on that port.

Once a port is determined to be PoE enabled and power is applied to it, it must continue to draw a minimum current. If it doesn't (device unplugged), the PSE removes network power and marks the port as non-PoE-enabled. The discovery process is then repeated until a PoE-enabled device is found.

10.4 What Are the Advantages to PoE?

PoE technology is driven by the need to install wireless access points in locations without AC power at a cost that is not prohibitive. Depending on the location of the access point, the cost of conduit, wire, and labor to deliver power to an access point was sometimes many times more than the cost of the access point hardware.

Even though wireless access points were the driving factors behind PoE, it now includes applications in many other areas including:

- Remote security cameras
- Data-entry terminals
- Remote displays
- Vending machines
- Gaming machines
- Point-of-sale terminals

- Thermostats
- Voice over Internet Protocol (VoIP)

Another huge potential application for PoE is the charging of battery-driven devices like cell phones, pagers, PDAs, and the like. These devices now all have an assortment of special cables and adapters. If you travel overseas, you sometimes need an adapter for every country you visit as the power line connector can vary from country to country. If your battery-powered devices were equipped with standard RJ-45 connections, every Ethernet connection (standard over the world) could potentially become a source for charging your device.

PoE also provides much needed flexibility to system managers. Built into PSE devices is the ability to manage the power delivery to PDs remotely. Using Simple Network Management Protocol (SNMP) functionality a device can be powered on or off remotely allowing a device to be rebooted from anywhere in the world.

10.5 How Do I Get Started with PoE?

There are two ways to begin using PoE. First, you could replace your existing switches with PoE-enabled switches, also called *inline power* devices. The PoE-enabled switches include a power supply connection that provides the source of the network power.

Alternatively you can begin by using a mini-span hub, also know as a "power hub." This device is inserted between your switch and the network devices. It provides PoE functionality for non-PoE-enabled devices. You simply disconnect your device from the switch and plug it into the hub and then connect the hub to the port on the switch. While a mini-span hub is a way to maintain your existing infrastructure, it has the disadvantage of adding to the wiring, power, and space requirements of your wiring closet.

Prior to starting work with PoE, surveying your existing wiring is mandatory. Are all your devices serviced by CAT5 cable? If not, you will want to upgrade your cable to CAT5. Second, is all your CAT5 wired properly? Many times, installers knowing that there were unused wire pairs in CAT5 cable will "double up" devices on a single cable. PoE will not work with nonstandard wiring.

You will also want to consider the need for a UPS system to backup the power delivered to your network. If you are providing critical services to your company, a UPS is probably another addition you will want to make to your wiring cabinet.

Finally, check the version of your SNMP management software. Make sure that your version supports PoE.

10.6 Resources

- **http://www.iol.unh.edu/consortiums/poe/** – Power over Ethernet Consortium website
- **http://www.ietf.org** – Internet Engineering Task Force (IETF) for documentation on Internet Drafts
- **http://www.ieee802.org/3/af** – IEEE Power over Ethernet web pages

11.0 Wireless Ethernet

Today it is simply expected that wherever you go—hotels, restaurants, airports, convention centers, schools, and even some churches—that you're going to find a Wi-Fi connection. The radio waves that transmit data in these applications make this technology different than wired Ethernet networks. Although the contents of the base Ethernet packet is the same as a wired packet, the terms, technology, procedures, and practices for operating a wireless Ethernet system are different.

11.1 A "Very" Short Technology Primer

IEEE standard 802.11 defines the wireless communication method used in today's wireless enterprise networks. Unfortunately, there is an alphabet soup of standards within this specification for us to sort through. Before 802.11 Part a (802.11a for short) was developed, a version with less throughput, 802.11b, was adopted by some companies. After those standards, there was 802.15.1, also known as Bluetooth, the security standard 802.11i, and others like 802.11g. Fortunately, most of this headache really belongs to the wireless equipment manufacturers. All we need to understand is the basic differences between the two standards most commonly used today, 802.11a and 802.11b. These two standards are summarized in Table 11-1.

Table 11-1. Comparison of 802.11 Part A and B

802.11a	802.11b
54 Mbps in 5-GHz band	11 Mbps in 2.4-GHz band
Shorter distances	More commercial applications
Higher throughput	Frequency conflicts with common devices like microwave ovens
Generally less interference	Can be lower cost than 802.11a
Less crowded frequency spectrum	Adequate throughput for email, web browsing

802.11a is generally applied to higher throughput applications like video, high bandwidth data, and some audio. It is also more noise immune and can more easily accommodate large numbers of users, which may bog down an 802.11b network. Of course, the greater speed, functionality, and noise immunity of 802.11a are obtained at a higher price point.

Both 802.11a and b are grouped in the wireless network domain as Wi-Fi networks. These networks span from 10 to 100 m. In addition to Wi-Fi, some people talk about wireless personal area networks (WPAN). These networks are for desktop applications like printer sharing, telephone handsets, and the like. WPAN networks usually work within the 0 to 10 m range. Large networks are known as wide area networks (WAN) and are used to link buildings together. In this chapter, we will focus on Wi-Fi, the most common networks for industrial applications.

When thinking about wireless communications, think about air as just another communications medium. With wired Ethernet, we use CAT5 cable to physically transfer data from one point to another. As we discussed in earlier chapters, we have some physical interface to the media and a software interface known as a media access controller (MAC). When we look at wireless, we have the same components. We have a medium, a physical interface, and some media access control. The medium is air. The physical interface is the hardware that converts the bits and bytes of the Ethernet message to a wireless signal. The MAC is the software/hardware layer that monitors the channel and controls access, monitors the channel for collisions, and schedules messages for transmission.

A few basic physical communications standards used in wireless communications include spread spectrum, orthogonal frequency division multiplexing (OFDM), and infrared. Of these the spread spectrum standard is the most interesting. In spread spectrum, data is transmitted using a number of channels. A part of the message is sent on one frequency, the device "hops" to the next channel, and then sends the next part of the message. The number of hops, the channels used, and the sequence account for some of the differences in the different wireless networks. In general, the more hops, the less opportunity there is for your entire message to get destroyed by outside interference.

No matter which physical standard is used, interference can disrupt the operation of your wireless system. Interference sources can include Bluetooth devices, cordless phones, and neighboring wireless LANs. As the number of wireless devices continue to grow, increasing attention needs to be made to these interference sources <u>before</u> your wireless system is deployed.

Another source of lost packets and lower bandwidth is attenuation. Attenuation is a decrease in the amplitude of the signal as it radiates out from the source. Signals are attenuated by everyday objects including walls, machinery, pillars, floors, and furniture. The range of any wireless LAN is a function of the number and type of objects between the transmitter and receiver. The completion of a site survey prior to deployment identifies the presence of attenuators and locates access points to obtain the greatest coverage through the facility.

Several vendors make site survey tools to assist vendors in the site survey process. These tools are designed to find sources of interference and attenuation so that access points can be located or aimed to get the greatest throughput.

11.2 Access Points

The typical low-cost access points available from your local retail computer store provide a shared wireless network where clients take turns transmitting their data. These access points are similar to an Ethernet hub in a traditional wired Ethernet network. Just like their wired cousin, a wireless LAN wants only one message at a time on the network. However, just like the wired version, the more users on the network, the more collisions, the more packet retransmissions, and the less throughput for everyone.

The more expensive wireless access points are more like Ethernet switches than hubs. In a wired switch, each node can transmit at any time and the switch resolves a lot of the collisions by retransmitting the message only to the selected destination, not to everyone on the switch. Since everyone can transmit simultaneously, there are fewer collisions and greater throughput. Many of these wireless switches attain the same type of simultaneous transmission and greater throughput by using directional antennas to aim signals at a particular node and using multiple channels. Besides simultaneous transmission and fewer collisions, these types of access points provide the network with much better range. Again, these wireless switches achieve this greater bandwidth at a significantly higher cost.

11.3 Mesh Networks

An alternative to expensive switched access points is a wireless mesh network. Mesh networks are networks of wireless nodes that can relay messages around the mesh to locate a specific node. These networks can easily accommodate sudden interference, are more mobile, and are more adaptable to a changing environment. Mesh networks can transfer messages in any number of relays through a varying node sequence. If a message is typically transferred from Node A to Node D through Nodes B and C, it will automatically transfer a message between A and D using Nodes B, G, M, and T if Node B suddenly loses access to node C.

The downside to mesh networks is the lack of common standards. Most mesh network systems are more proprietary than open with routing software particular to their brand of mesh network. Users should think long and hard before committing their future to a single source system.

11.4 Security

Wireless security is such an important and complex topic that it really warrants an entire book. Instead, this section will attempt to provide a basic explanation of the problem and how to deal with it.

Because radio waves propagate through walls and outside your physical space, wireless LANs pose a threat to the integrity of your IT system and the operation of your entire enterprise. For example, the transmissions of a wireless LAN can be passively monitored for a long distance from your facility using directive antennas. If you don't implement the minimum standard security mechanisms built into your wireless devices, outside eavesdroppers can read emails and access files sent between users.

There are any number of methods individuals may use your wireless LAN to harm your facility. The most common include:

- Service set identifier (SSID) sniffing
- WEP encryption key recovery attacks
- ARP poisoning
- MAC spoofing
- Access point password attacks
- Wireless end-user station attacks
- Rogue AP attacks
- Denial of service (DoS) attacks
- Planned cordless phone interference

The most common method to prevent these attacks and others like them is to activate the Wired Equivalent Privacy (WEP) security included in most wireless devices. WEP encrypts the body of each data frame and is designed to prevent unauthorized users from detecting email addresses, user names, passwords, and viewing sensitive documents. Unfortunately, weaknesses in the WEP security system have been detected and hackers can sometimes break into WEP-protected systems in as little as 24 hours.

One of the weaknesses of the current WEP system is the encryption keys used by the wireless transmitter and receiver. Both ends must know the encryption key. But because management of these keys is such a headache for the wireless systems administrators, these keys are hardly ever changed. Without frequent rekeying of a wireless system, hackers can get months and months to work at discovering the keys and cracking the WEP security.

The IEEE 802.11 Working Group addressed the weaknesses of WEP by creating the IEEE 802.11i standard, which includes not only extremely reliable encryption of data packets but also the rotation of encryption keys. The key rotation feature alone will go a long way toward discouraging all but the most determined hackers.

Another security consideration that is often neglected is wireless access to your enterprise network over public wireless systems. Even though WPA and 802.11i deal with the wireless security your data must still be transferred over a wired Ethernet system from the access point outside your facility. To protect your data over the wired systems, companies should ensure that users have virtual private network (VPN) client software. A VPN encrypts data all the way from your remote user to the corporate VPN server.

11.5 The Advantages

The advantages of wireless LANs are numerous. There are, of course, economic factors. Components such as access points, wireless adapters and other network hardware are continually dropping in price. Installation is faster and less expensive without the requirement to run cabling and install a wall out for each user. Wireless access is faster as the system is immediately operable once the Access Point is configured. And if your

Access Point uses Power over Ethernet (PoE), no power cables are required.

Over the long term, the ability to reconfigure your office without reconfiguration of your network, the lack of cabling to be mistakenly cut and the higher reliability of wireless LANs are significant advantages of a wireless system. Some argue that the long-term cost of ownership of a wireless LAN is less than its wireless counterpart. They argue that not only is the overall per point cost lower but the productivity advantages of a wireless system over a wired system are significant.

12.0 Advanced Hardware Topics

If you work in any kind of manufacturing facility or industrial environment, you've probably heard about time synchronization, dual Ethernet devices, redundant Ethernet networks, and ring technology. This chapter provides a short introduction to each of those advanced Ethernet topics.

12.1 Time Synchronization

You may not realize this, but on Sunday morning when you are reading your online newspaper, the Ethernet packets that deliver the content are traveling different routes to reach you. The routers and switches between you and the newspaper website are automatically finding the best route at that particular moment for the next page of the article. Because each packet takes a different path, the time it takes a packet to move from the news website to your computer varies.

Unless you are an amazingly fast reader, it is likely that the variances in packet delivery time don't matter to you. But in a number of telecommunications, industrial, military, and power distribution applications, those tiny discrepancies are unacceptable. Imagine trying to start a group of motors that pull a long sheet of paper through a papermaking machine. If those motors are not precisely aligned and do not maintain that alignment, that web of paper quickly becomes confetti.

There is no illusion that standard, off-the-shelf Ethernet can be used to achieve any kind of real-time performance or device synchronization. There are three common problems. The first, as discussed previously, is that packets are routed differently and delivery times vary. The second problem is that Ethernet devices must decode the TCP/IP stack protocol layers to obtain the application message. On Ethernet platforms with different operating systems, with different TCP/IP stacks, and different processor speeds, there is no consistent decode time. Even if Ethernet could consistently deliver a message with consistent repeatability, there is no guarantee that the message would be processed in exactly the same time. The third problem with achieving near real-time performance is the variability of clocks on different platforms. Even two clocks on the same platform from the same manufacturer will vary due to age, temperature, and other factors to make synchronized applications unworkable. Network Time Protocol (NTP) and other time protocols solve that problem for non-precise timing applications, but are not nearly accurate enough for the highly synchronized applications in telecommunications, power, and complex manufacturing systems.

IEEE 1588 is the technology that solves these problems and allows remote Ethernet devices in these applications to synchronize to a common time clock and coordinate operation very precisely. It solves the problems by providing:

- Automatic evaluation of all the Ethernet device clocks to designate the most accurate and reliable clock as the system master clock
- Automatic replacement of the master clock if it becomes less reliable
- Continuous synchronization of all device clocks to the master clock
- Minimal bandwidth consumption and processing power
- Limited configuration and setup on deployment
- A mechanism for devices to direct applications to execute application activity at a precise time, often in the submillisecond range

IEEE 1588 provides these benefits by using the Precision Time Protocol (PTP) to deliver fault-tolerant synchronization of the clocks in a network of remote Ethernet devices. PTP defines the value ranges for the standard set of clock characteristics and a best master clock (BMC) algorithm. The BMC algorithm determines which clock is the highest quality clock using a combination of a priority value, the source of the clock operation (free running, GPS, etc.), its accuracy, and its variance over time. All other clocks in the network (slave clocks) are synchronized to the BMC (the grandmaster clock). If the BMC is removed from the network or is determined by the BMC algorithm to no longer be the highest quality clock, the algorithm then selects a new BMC and all other clocks synchronize to it. No administrator input is needed for this adjustment because the algorithm automatically selects the current best clock to use as the BMC.

The process of sending and receiving synchronization packets allows the slave clocks to accurately measure and offset between the slave's own clock and the master clock. Standard methods of clock adjustment implementation are not outlined by IEEE 1588; it only provides a standard protocol for the exchange of messages between clocks. The benefit of this is that clocks from different manufactures are still able to synchronize with each other.

The quality of the Ethernet-based IEEE 1588 system and the way it is set up can affect the quality of synchronization. To set up the best synchronized system, one must make a tradeoff between exactness of synchronization, cost, and distance needs. For low-speed events that do not depend on time, a standard NTP synchronization over the Internet, which allows for millisecond-level synchronization, will suffice. For applications needing sub-microsecond synchronization in geographically distributed systems, IEEE 1588 can be implemented. More accurate synchronization is achieved by using IEEE 1588-compatible Ethernet Physical Interfaces (PHYs).

12.2 Dual Ethernet Devices

An increasing number of industrial devices are being produced with two Ethernet ports (Figure 12-1). These devices are used in applications where Ethernet devices are deployed in a daisy-chained fashion: the first port is connected to the previous unit in the chain, the second port is connected to the next unit in the chain, and so forth.

Figure 12-1. A Dual Ethernet Device

A tiny Ethernet switch makes this type of device possible. Most of the switches you use every day are based on a single piece of silicon (chip) that has all the functionality built right into it. One of the benefits of the explosion of Ethernet devices over the last 20 years is that these components are now smaller and much less expensive. It reached the point recently where these chips are so inexpensive that ordinary I/O devices, motor drives, and other everyday industrial Ethernet devices can have a switch embedded in them.

Having an embedded switch and a dual Ethernet device presents all sorts of possibilities for integrating Ethernet devices into networks and for configuring those networks. For example, dual Ethernet ports can be configured to provide:

- **Daisy-Chained, Trunk, and Drop configuration** – The embedded switch is configured to send any message received on one of its ports out the other port. Messages then can be passed from device to device. This kind of configuration is perfect for systems where devices are organized in a line one after another like you might find on a long conveyor line. In this configuration, Ethernet competes directly with protocols like DeviceNet and PROFIBUS, which are most often used for linear, trunk-and-drop type applications.
- **Ring-Network Configuration** – When the outgoing port of the last node in the link is connected back to the incoming port of the first node in the network, the devices form a ring. (See the next section.)
- **Dual-Network Device Configuration** – This is configuring the switch to separate its ports so that each port is on a different network. This configuration is used in applications where a factory floor network is attached to one port and an enterprise network is attached to the

other port. The device can then choose what information to share between the factory and the enterprise.

- **Dual-Port Device Configuration** – The two ports on the switch could also be configured as two separate Ethernet ports on the same network with two different physical network addresses. This enables the device to "pretend" to be two different devices on the same network.

12.3 Device Ring and Redundancy

As discussed in the last section, connecting the last node on the Ethernet link to the first node, creates an Ethernet ring (Figure 12-2). This network configuration presents both advantages and challenges for industrial applications.

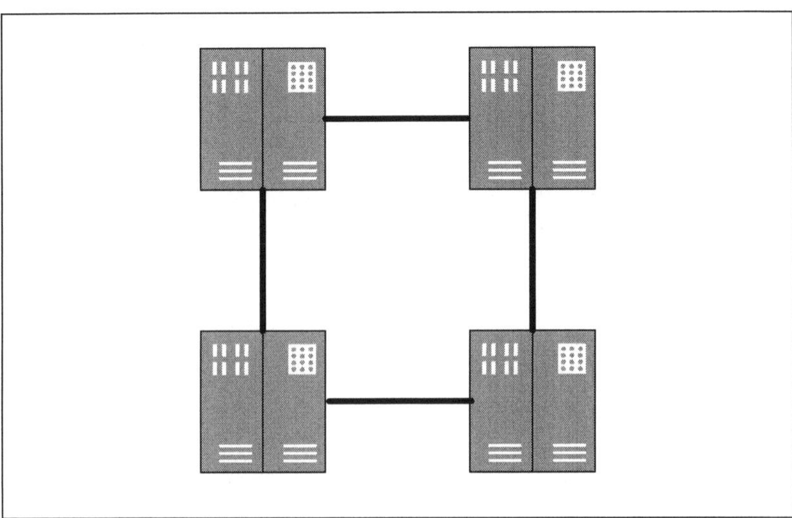

Figure 12-2. Simple Device-Level Ethernet Ring

An advantage to this type of network configuration is a certain amount of redundancy. This is very important in applications where a cable failure (Figure 12-3) could result in disastrous consequences for the production process, plant personnel, or even the general public. This type of media failure can be detected when one node fails to receive the low-level connection messages that Ethernet sends between nodes.

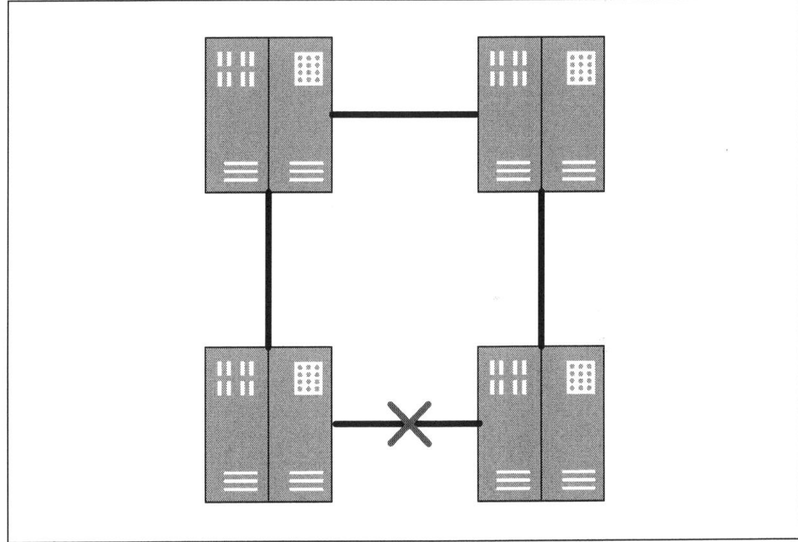

Figure 12-3. Ethernet Device Ring with Cable Failure

In some ring networks, a control node (Figure 12-4) provides redundancy. The control node initiates all packets from its outgoing port and receives them on its incoming port. If a transmitted packet does not arrive at the incoming port, the control node assumes that there is a cable break and it sends packets out from both ports from that point on.

A key consideration for ring networks is how fast the system recovers from a link failure. In standard, commercial systems this can be as much as 60 seconds. In an industrial system serious damage can happen quickly, so systems are specifically designed to detect cable failure in as little as a few milliseconds.

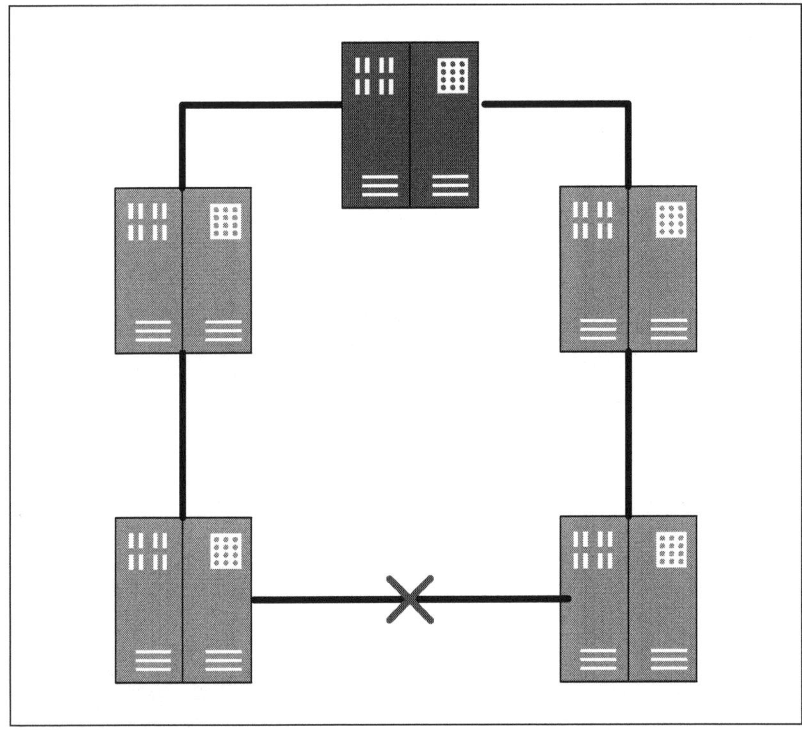

Figure 12-4. Device-Level Ring with Controlling Node

Without a control node in the system, there is no beginning and no end to the network; the packet can theoretically be passed from one node to the next forever. In standard Ethernet, there is a packet lifetime after which that packet is not passed to the next node. The lifetime limitation could be used in an Ethernet ring, but a lot of bandwidth would be wasted by sending packets around and around until they die of old age. In practice, one way to make this strategy work is for each node to determine if the arriving packet is the packet it sent out on its other port. If the packet is received, the device knows to discard it as it has traversed the entire network.

Another strategy is to designate part of the network as a "secondary" link (Figure 12-5). Messages are not sent on the secondary link unless a break is detected in one of the primary links. There are two advantages to this strategy: (1) packets die on reaching the end of the link and (2) the secondary link can quickly be activated.

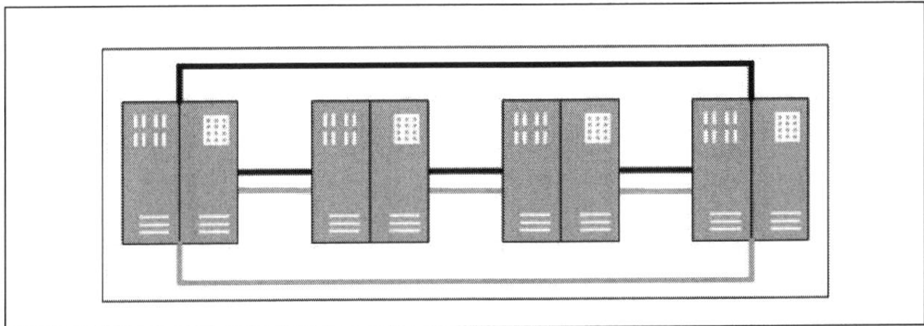

Figure 12-5. Ring Network with Standby Secondary Link

Trunking is another mechanism for creating redundant systems. In a trunking network, each device-to-device link is connected by multiple cables. Should one cable fail, there is another cable to keep that link alive. These systems provide the fastest switchover on failure. Trunking, due to the extra wiring, the extra cabling, and so forth, is the most expensive solution and is implemented in only the most critical of production processes.

12.4 Summary

Vendors of Ethernet devices are constantly shrinking the size of Ethernet components and adding new functionality. Switches have evolved from large, expensive devices to small, integrated circuit components that are included within computer processors. This evolution has spawned all sorts of new applications like the redundant Ethernet, synchronized Ethernet, and Ethernet rings discussed in this chapter. More development is surely on the horizon. Progress will continue. It will be interesting to see the next step in the Ethernet evolution.

13.0 The Internet of Things

I can't think of any term in the history of technology that is so misused, so misunderstood, and so overhyped as the Internet of Things (IoT) or the Industrial Internet of Things (IIoT). So what is IoT, and how does it differ from traditional Ethernet networking?

IoT differs from traditional computing in several ways. Unlike most of the applications you run today, these applications transfer small data packets, usually infrequently, and are designed to accommodate hundreds of thousands of devices. Also, instead of you communicating over an Ethernet network, machines or sensors are talking to other machines or sensors.

All the largest technology companies are aggressively moving into the IoT with sophisticated, complex, and highly functional cloud-based service platforms to solve customer problems. This chapter describes how Microsoft, Amazon, and Oracle are approaching IoT, and specifically IIoT.

13.1 Microsoft and the IoT

Microsoft builds its approach to the IIoT around Microsoft Azure, its cloud application platform. Azure is a massive collection of capabilities; there are compute services (which run a lot of your mobile phone applications), traditional data management services (SQL services being an example familiar to many people), big data analysis, mobile services, backup site management, streaming analytics, and much more.

Though most of these services can be important to industrial applications, Microsoft is hoping that manufacturers will want to connect their machines to Azure to take advantage of the analytics services, either the HDInsight big data services or the real-time analytics available in the Stream Analytics service. These services are designed to look at data and provide insights into reliability, product quality, maintainability, and other system indicators.

Microsoft provides several ways for factory floor devices to get data into Azure, but they all revolve around a virtual device called the *IoT Hub* (see Figure 13-1). The IoT Hub is a fully managed service that provides reliable device-to-cloud messaging for any size application. It can scale from a few devices with few data transmissions to hundreds or thousands of devices with many transmissions.

Factory-floor, Ethernet-enabled devices that can support IoT Hub protocols (MQTT, AMQP, and HTTPS) can directly send their data to the IoT Hub. Other Ethernet-enabled devices that do not support those protocols might be able to use Microsoft's protocol gateway that can connect other protocols to the IoT Hub. Non-Ethernet-enabled devices must use some sort of field gateway to move their data onto Ethernet in order to get connected to an IoT Hub.

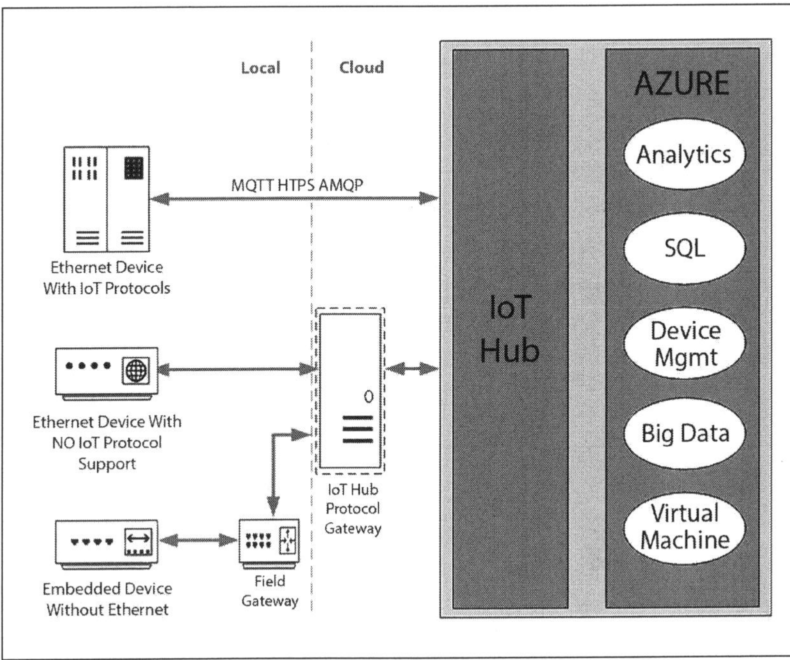

Figure 13-1. Microsoft Azure and the Internet of Things

13.2 Amazon and the IoT

Amazon first began offering web infrastructure support to businesses in 2006 as a better way for business to host their computing resources. Instead of building and maintaining installations to house servers, businesses could use Amazon's servers and not have the expense of maintaining them themselves. Additionally, they would benefit from the ability to scale up or down as compute resource requirements varied.

The web infrastructure business has now evolved into Amazon's Web Services (AWS) business. The IoT infrastructure support AWS provides to businesses accounts for a significant portion of their operating profit. AWS provides capabilities that are similar to those provided by Microsoft Azure, but it is organized differently and uses an entirely different mechanism to ingest the data that is processed by its IoT services.

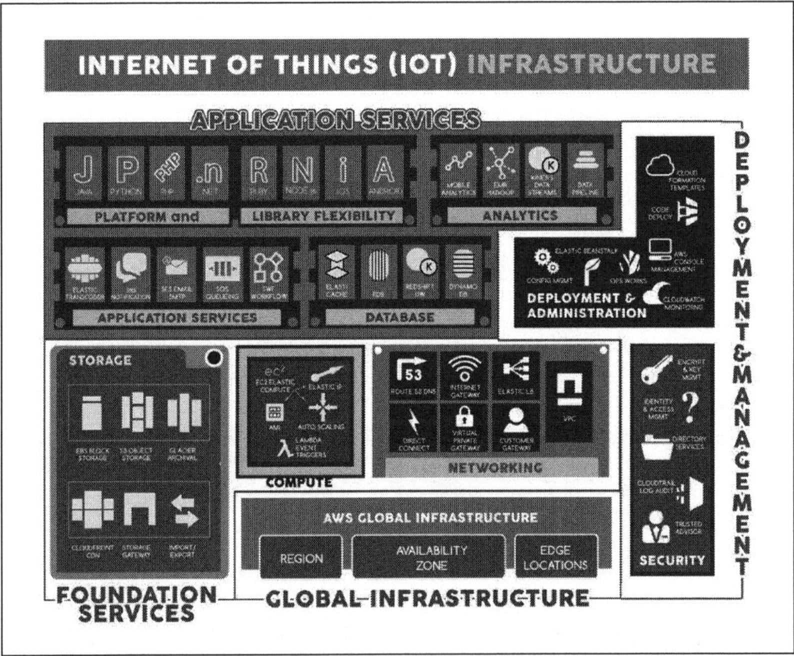

Figure 13-2. Amazon and the Internet of Things

Amazon describes its AWS cloud services (see Figure 13-2. Amazon and the Internet of Things) in terms of building blocks:

- **Global Infrastructure** – The building block responsible for the global network of servers and computing resources located around the world.
- **Security** – The mechanisms built into AWS that provide authentication, authorization, and encryption of data maintained by AWS.
- **Foundation Services** – The basic services needed by every IIoT application composed of compute services

(the ability to provide compute resources as needed), storage mechanisms, and networking services.

- **Application Services** – Services available for use by applications including email, workflow, language tools, databases, and advanced analytics, which includes big data services.
- **Deployment and Management** – Services that provide managers with the ability to monitor the operation of their AWS services.

13.3 Oracle and the IoT

Where Amazon and Microsoft overwhelm the user with IoT services and complexity, Oracle has created an IoT cloud service that is simpler, easier to understand, and likely much easier to implement. That is not to say that Oracle's Cloud Service is any less functional. It has all the key features that the vast majority of IoT customers might want.

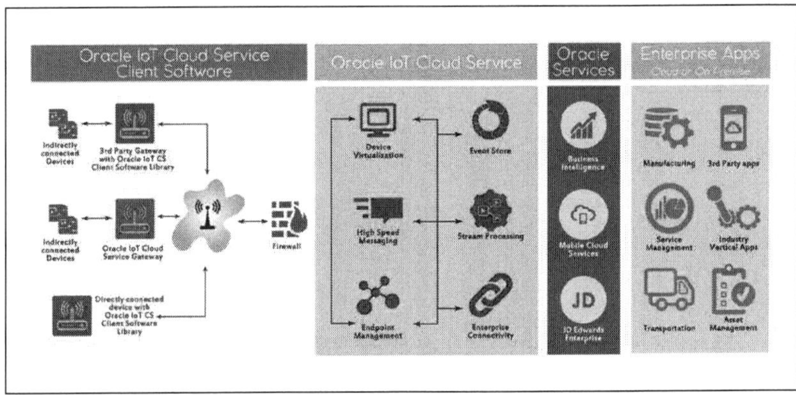

Figure 13-3. Oracle and the Internet of Things

Figure 13-3 illustrates the architecture of the Oracle IoT Cloud Service. The architecture of how IoT data is ingested is very similar to the architecture used by both Microsoft and Amazon. There is a mechanism for ingestion of data from directly connected devices. There is a virtual gateway in the cloud for devices that can send Ethernet messages, and there is a provision for third-party gateways for devices that are not Ethernet enabled.

The technology that Oracle uses for ingestion of IoT data is different from that used by Amazon and Microsoft. Oracle use the RESTful architecture (also called *RESTful interface*). RESTful is a very flexible architecture, usually built on top of HTTP, for client devices to make follow-up requests of server devices using well-defined and simple processes. It is resource-centric instead of function-centric. In the RESTful architecture, a server is viewed as a set of resources, nouns if you will, that can be operated on by a simple set of verbs like GET, POST, UPDATE, and the like. This is the opposite of a function-centric technology where there are a set of fixed functions that are available to the client. Both the client and server need software changes in a function-centric architecture to add new capabilities. That's not true using a RESTful architecture. In RESTful, capabilities can be easily added by adding nouns, making it a much more flexible mechanism for retrieving resources than the limited, function-centric technology.

Once data is ingested, it is stored in the Oracle IoT Event Store where it can be accessed by Oracle's Business Intelligence services, Visualization services (charting, graphing, etc.), Mobile services, and enterprise applications offered by Oracle, third-party vendors, or individual users.

13.4 Summary

Microsoft, Amazon, and Oracle are all making great efforts to provide data services for IoT data. They see this as a competitive and lucrative business opportunity. This chapter described the IoT offerings from each of these vendors. The offerings are strikingly similar, providing a lot of the same data capabilities. All vendors provide similar mechanisms for getting data into their systems. Once captured, the data is available for visualization (display), analysis, archiving, or as input to custom applications. The specific services that operate on the IoT data are quite similar but their implementation, ease-of-use, and business model vary quite a bit. For example, each vendor has an analytics service, but how the services work, their capabilities, and the pricing models are very different. All these companies, and others, will all continue to add more services and more capabilities to more easily capture more data. This is just the beginning of the evolution of the Internet of Things.

14.0 Factory Floor/ Enterprise Communications

If you have paid any attention to factory automation over the last few years, you have noticed the ever-increasing emphasis on connecting the factory floor to the enterprise. There are many good reasons for this. Some of the reasons are internal: efficiency, productivity, higher quality, and the like. Others are driven by external requirements. Large customers, such as the Walmarts of the world, are demanding higher levels of integration with their suppliers. Regulators are increasing their demands for manufacturers to report on their production processes. Corporate attorneys are "suggesting" that manufacturers archive more data about their production processes.

It wasn't always like this. In the old days (10 years ago?), the production department was a completely separate entity from the rest of the corporation. There was little-to-no electronic data transfer between the production machines and the company's business systems. Production was a black box. Labor and raw materials went in one end, and finished product came out the other end. Most of the communication was carried out using paper: paper production reports, paper inventory levels, paper raw material usage, paper quality reports, etc. People keyed this information into business systems that used sales data to order raw materials and adjust production levels for the next production cycle—again using paper systems.

Today, the aim is for instantaneous closed-loop communication. As units of product are consumed in the field, that information gets reported back to the machine that made it. The production machine checks its raw material inventory levels and on-hand finished product, and then schedules more production. It automatically transmits orders for any raw materials it needs to supplier machines. All automatic. All without human intervention.

That is the plan anyway. In practice, it is pretty hard to get there. We do not have the luxury of ripping out all the production machines and replacing them with new, fully integrated machines with high-speed communication mechanisms. Instead, we have to do piecemeal implementations: upgrading and replacing systems one by one as time and funds allow. It is a marathon, not a sprint, to the goal of fully automated systems.

There are many factors impeding progress on our path to fully integrated production systems. Security, of course, is key—the more integrated and connected your production process is, the more vulnerable you are to mischief and worse. Another factor is the difficulty of replacing fully capitalized and functional systems that are well understood and perfectly operational but lack the integration required for tomorrow's manufacturing vision. Yet another is the mutual lack of understanding of IT integration by today's control engineers and manufacturing by today's IT people.

14.1 Tight versus Loosely-Coupled Systems

The distinction that many people from manufacturing and IT miss is that there is a key distinction between the systems on the factory floor and the enterprise system. This is the difference between what is called *loosely-coupled* systems and *tightly-coupled* systems. These are not new concepts, but I don't think they have been examined in the light of the current trend towards the integration of factory floor and enterprise systems.

Factory floor systems can be labeled tightly-coupled. Systems that use PROFIBUS, PROFINET IO, DeviceNet, EtherNet/IP, or any Modbus version have a strict architecture. These are really just I/O producers and consumers, despite what some folks at the trade associations might want you to believe. Let's look at the main characteristics of tightly-coupled systems:

- **A Strictly Defined Communication Model** – The communication between these systems is inflexible, tightly regulated, and as deterministic as the communication platforms allow.
- **A Strictly Defined Data Model** – The data model (really an I/O model for most of these systems) is predefined, limited, and inflexible.
- **Strictly Defined Data Types** – The data types transported by these systems are limited, predefined, and supported by both sides. There is no ability to send data in an open and universal format.

We could look at any of the factory floor protocols, but let's take EtherNet/IP as an example. EtherNet/IP has a strictly defined communication model. A scanner uses a precise communications model in communicating with its adapters. The adapters are preconfigured: all data exchanged is predefined, and nothing changes without human intervention. The data exchanged is part of the adapter's predefined object model, and the data is formatted in a way supported by both the scanner and the adapter.

Tightly-coupled systems provide much needed, well-defined functionality in a highly specific domain. Expanding operation to other domains or trying to provide more general operation is difficult. Making more generic data and functionality available requires significant programming resources that results in a very inflexible interface.

That is why tightly-coupled systems are wrong for enterprise communications. Can they be made to work for a specific application? Yes. But to get there requires a tremendous effort and results in a difficult-to-maintain, inflexible system that is extremely fragile. These systems are difficult to maintain as any small modification anywhere along the line can cause a failure.

Loosely-coupled systems, on the other hand, provide exactly the right kind of interface for enterprise communications. Loosely-coupled systems decouple the platform from the data, decouple the data from the data model, and provide a much more dynamic mechanism for moving data. Loosely-coupled systems have these characteristics:

- **A Widely Used, Standards-Based Transport Layer** – Messages are transported in loosely-coupled systems with open, widely-implemented, highly flexible transports layers: TCP and HTTP.
- **An Open, Platform-Independent Data Encoding** – Data is encoded using an open standard data encoding like XML that can be processed by any computer platform.

- **A Highly Extensible Operating Interface** – The interface between loosely-coupled systems is flexible and extensible. SOAP (the Simple Object Access Protocol, which is rapidly being replaced by REST) is the main interface, and it provides a highly flexible mechanism for messaging between loosely-coupled systems.

Essentially, what I've described here is web services. The web services architecture is the backbone of everything we do on the Internet. It is extensible, flexible, and platform independent—all required for the ever-expanding Internet.

The challenge is to how to best migrate the tightly-coupled factory floor architectures to the loosely-coupled web services architecture of the Internet. It is difficult to migrate today's technologies for transferring I/O data like Modbus TCP, EtherNet/IP and PROFINET IO. It can be done, but it often results in brittle systems that require too much support and cost too much time and money. Integrating these technologies with loosely-coupled enterprise technologies takes massive amounts of human and computing resources. In the process, we lose lots of important metadata, we lose resolution, and we create security concerns. These factory floor systems were not designed to be highly secure. Using the factory floor protocols for enterprise data collections creates systems that are a house of cards.

Because of the discontinuity between the factory floor systems and the enterprise systems, opportunities to mine the factory floor for quality data, interrogate and build databases of maintenance data, feed dashboard reporting systems, gather historical data, and feed enterprise analytic systems are lost. Opportunities to improve maintenance procedures, reduce downtime, and compare performance at various plants, lines, and cells across the enterprise are all lost.

14.2 OPC UA for Factory-Enterprise Communications

The solution is OPC Unified Architecture (UA), a new architecture for moving information between manufacturing and the enterprise. OPC UA can live in factory floor world and the enterprise world.

OPC UA is about reliability, security, and most of all, easily modeling *objects* and making those objects available around the plant floor, to enterprise applications, and throughout the corporation. The idea behind OPC UA is infinitely broader than anything most of us have ever thought about before.

It all starts with an object. An object that could be as simple as a single piece of data or as sophisticated as a process, a system, or an entire plant. It might also be a combination of data values, metadata, and relationships. For example, let's consider a dual loop controller: the dual loop controller object would relate variables for the set points and actual values for each loop. Those variables would reference other variables that contain metadata like the temperature units, high and low set points, and text descriptions. The object might also make subscriptions available to get notifications on changes to the data or the metadata for that data value. A client accessing that one object can get as little data as it wants (single data value) or an extremely rich set of information that describes that controller and its operation in great detail.

OPC UA is, like its factory floor cousins, composed of a client and a server. The client device requests information. The server device provides it. But, the way that the OPC UA server processes the information is much more sophisticated than the process performed by an EtherNet/IP, Modbus TCP, or PROFINET IO server. An OPC UA server models data, information, processes, and systems as objects and presents those objects to clients in ways that are useful to vastly different types of client applications. And better yet, the OPC UA server provides sophisticated services that the client can use, like the Discovery Service, a service that locates available OPC UA devices.

14.3 Ten Things to Know about OPC UA

OPC UA is the future and the perfect technology to bridge the chasm between loosely- and tightly-coupled systems. Here are 10 things you need to understand about OPC UA:

1. **OPC UA is not a protocol.**

   It is a common misconception that OPC UA is just another protocol. That could not be further from the truth.

   A computer protocol is a set of rules that govern the transfer of data from one computer to another. Even though OPC UA also specifies the rules for communication between computers, its vision is more than just moving some arbitrary data from one computer to another. OPC UA is about complete interoperability.

   OPC UA is an architecture that systematizes how to model data, model systems, model machines, and model entire plants. You can model anything in OPC UA. OPC UA is a systems architecture that promotes interoperability between all types of systems in various kinds of applications.

2. **OPC UA is the successor to OPC (now referred to as *OPC Classic*).**

OPC UA solves the deficiencies and limitations of OPC Classic, a technology built on the now obsolete Microsoft COM. In today's world, we need to move data between all sorts of embedded devices, some with specialized real-time operating systems and software, and enterprise/Internet systems. OPC Classic was never designed for that.

OPC UA is the first communication technology built specifically to live in that "no man's land" where data must traverse firewalls, specialized platforms, and security barriers to arrive at a place where that data can be turned into information.

3. **OPC UA supports the client-server architecture.**

We are all familiar with technologies that have a superior/subordinate relationship, often with one master to many slaves. That is not true of OPC UA clients and servers. In OPC UA, a slave can be configured to accept connections with one, two, or any number of clients. A client device can connect and access the data in any number of servers. It is much more of a peer relationship in OPC UA, though, like other technologies, servers simply respond to requests from clients and never initiate communications. In practice, many devices are being built to easily support peer relationships implementing both client and server functionality.

Another unusual and interesting aspect of the client/server relationship is that in OPC UA, a server device can allow a client to dynamically discover what level of interoperability it supports, what services it offers, what transports are available, what security levels are supported, and even the type definitions for data types and object types. These characteristics make an OPC

UA server much more sophisticated than the servers for many of the technologies you have worked with in the past.

4. **OPC UA is a platform-independent and extremely scalable technology.**

   Unlike OPC Classic, OPC UA is designed from the ground up to be platform independent. The only requirements for OPC UA are Ethernet and a mechanism to know the current date/time. OPC UA is being deployed to everything from small chips with less than 64K of code space to large workstations with gigabytes of RAM.

   All the components of OPC UA are designed to be scalable, including security, transports, the information model, and its communication model. Several security models are available that support the level of security appropriate for the device's resources and processor bandwidth.

5. **OPC UA integrates well with IT systems.**

   OPC UA servers can support the transports used in many traditional IT-type applications. Servers can connect with these IT applications using SOAP or HTTP (the foundation of the data communication used by the World Wide Web).

   OPC UA servers can also support XML encoding, the encoding scheme used by many IT-type applications. It is likely that most servers in the factory floor automation space will not support XML encoding due to the large amount of resources required to decode and encode XML. However, many servers in that space will support OPC UA Secure Conversation, a more efficient binary encoding that uses fewer resources.

6. **OPC UA provides a sophisticated address space model.**

   The address space model for OPC UA is more sophisticated than EtherNet/IP, PROFINET IO, Modbus, or any of the industrial or building automation protocols. The fundamental component of an OPC UA address space is an element called a *node*. A node is described by its attributes (a set of characteristics) and interconnected to other nodes by its references or relationships with other nodes.

7. **OPC UA provides a true information model.**

   The ability of an object node to have references to other object nodes that further reference other object nodes to an unlimited degree, provides the capability to form hierarchical relationships that represent systems, processes, and information—an *information model*.

   An information model is nothing more than a logical representation of a physical process. An information model can represent something as tiny as a screw, a component of a process like a pump, or something as complex and large as an entire filling machine. The information model is simply a well-defined structure of information devoid of any details on how to access process variables, metadata, or anything else contained within it.

   Many trade groups—including many in the oil and gas industry, the building automation industry, and PLC standards organizations, and others—are using the information model capabilities of OPC UA to define information models for their application domains. They are using OPC UA for the standard transports, security, and access to their data models.

8. **OPC UA extends factory floor communications.**

Instead of a factory floor protocol, you could say OPC UA is web services for automation systems, that it is SOA for automation systems. SOA is basically the same thing as web services. That is fine if you are an IT guy (or gal) and you understand those terms. You have some context.

But if you're a plant floor guy, it is likely that even though you use web services (it is the plumbing for the Internet). And it is just as likely you may also say, "Why do we need another protocol? Modbus TCP, EtherNet/IP, and PROFINET IO work just fine." The answer is that OPC UA is not like EtherNet/IP, PROFINET IO, or Modbus TCP. It is a completely new paradigm for plant floor communications. It is like trying to explain EtherNet/IP to a PLC programmer in 1982. With nothing to compare it to, it is impossible to understand.

OPC UA lives in parallel with these technologies. It doesn't replace them. It extends them by bringing in new functionality, creating new use cases, and driving new applications. In the end, it increases productivity, enhances quality, and lowers costs by providing not only more data, but the right kind of data to the production, maintenance, and IT systems that need it when they need it.

9. **OPC UA is a certifiable standard.**

   Like many other technologies, there is a process to validate that an OPC UA device conforms to the standard. And like many other technologies, there is documentation to certify that devices pass the OPC UA certification test suite. A successful compliance test results in an electronic test certificate being transmitted to the device. Client devices can then access the device and get the electronic certificate documenting its status as a certified OPC UA device.

10. **OPC UA is still a developing technology.**

    There is a technology adoption life cycle, and OPC UA is following that cycle. The first systems were available just a few years ago. Like any other technology adoption, there are the innovators, closely followed by the early adopters. If the technology is successful, the next group, the early majority adopters join in and then the technology reaches its peak adoption.

    OPC UA is the preferred communications protocol of numerous trade associations, of major vendors (such as SAP SE, one of the world's leading IT companies), and is the core for the entire German Industry 4.0 effort (the German government, industry, and educational systems combined effort to develop new manufacturing technology).

14.4 Reference

For a free book on OPC UA written by the author of this book, go to the web page listed below and leave a message asking for the book *OPC UA: The Everyman's Guide to OPC Unified Architecture* by John Rinaldi: www.rtaautomation.com/contact-us/.

14.5 Summary

There is much to gain by integrating the factory floor and the enterprise. There are efficiencies, flexibility, and vast amounts of data that can be mined to improve factory floor processes. For example, the vast majority of energy usage in a factory is from its motors but there is often no ability to access data on which motors consume the most energy, which are efficient, and which are inefficient. Enormous savings are available if this sort of information can be extracted from the factory floor.

But combining factory floor systems with the enterprise is challenging. The cultures, systems, and technologies are very different. Enterprise systems are flexible, open, and loosely coupled while many factory floor systems are closed, proprietary, and tightly coupled. OPC UA is a new technology which may be the bridge between the factory floor and the enterprise. OPC UA offers a unique set of capabilities that may unlock the data and information on the factory floor and yield the kinds of savings described above.

15.0 The Alphabet Soup of the Internet of Things

Not all Internet of Things (IoT) applications are being implemented using Microsoft, Amazon, Oracle, or other major company cloud services. A large number of users are building IoT applications using one of the many other protocols that can move data between field devices and applications in the enterprise or in the cloud. This chapter describes some of the Ethernet protocols that are being used to build these kinds of IoT applications.

15.1 XML

The Extensible Markup Language, known as XML, is becoming increasingly important in the world of industrial automation and the Internet of Things (IoT). It is a perfect storm of functionality and requirements. XML is a data language that communicates by sending files containing ASCII characters from one system to another.

XML is verbose—there is no getting around it. It is also simple and human readable. Whether one likes its verboseness and excessive bandwidth consumption or not, XML is the standard IT folks use worldwide. All of the standard offerings from Microsoft—like Word and Excel—are XML-based. And it's those same IT standards that are being pushed down to the factory floor.

What Is XML?

XML is a meta-markup language. That means that data in an XML document is surrounded by text markup that assigns tags to the data values. Each data value, when taken together with its distinguishing tag name, is an XML element—the basic defining unit of an XML document. An entire collection of elements forms the XML document.

An element is formed with a start-tag, an ASCII string, and an end-tag. All tags are enclosed in angle brackets like this: <…tag…>. End-tags signify that they are end-tags by preceding the tag name with a slash. Below are examples of well-formed XML elements:

<name> Emily Warden <\name>

<sentence> Where is the family dog? <\sentence>

<temperature> 22.53 <\temperature>

While the names for XML elements have few restrictions, XML documents follow a very specific and strict grammar. The grammar specifies where XML elements can be placed, how child elements are specified, how child elements are associated with parent elements, and how attributes are attached to elements.

The XML elements allowed in a particular application are defined in an XML Schema. An XML Schema defines all of the valid elements in a document and allows a generic parser to determine if an XML document is well-formed for a particular application. A document can be well-formed for one application (matching the chemical composition XML Schema) while being invalid for another application (not matching the XML Schema for a court case). Non-well-formed documents are ignored by a receiver.

How Is XML Used?

XML documents are standard text documents that can be created and edited in any text editor, or in a word processing program such as Microsoft Word. Once an XML document exists, it can be transported in any number of ways from a sender to a receiver. In many cases, a device is triggered to transmit an XML document by simply entering the URL for the XML document in a standard browser.

XML is a good choice for non-control data, especially when that data needs to be sent to different types of applications across various platforms. XML is essentially the lowest common denominator for transferring data. Anything that can decode an ASCII character can parse an XML file. XML is the most prevalent and highly integrated technology for moving data between two non-heterogeneous systems. There is no other technology that is supported across so many platforms and applications. XML parsers are built into browsers, databases, and many of the tools used to construct monitoring and data archival applications.

Summary

XML is a data encoding option for OPC UA. OPC UA clients can receive transmissions from OPC UA slaves where the data is encoded as XML. This option makes OPC UA attractive to many web service applications that already use XML to communicate with other IT applications.

XML encoding is also used in technologies like MTConnect, a device-information model that specifies the organization and contents of a machine tooling system. The entire MTConnect technology is based on XML including the Streams Information Model, an XML file that transfers values about a tooling system.

XML is perfect for IoT applications that send data to enterprise applications. Almost all Microsoft and enterprise applications can easily ingest XML data. The deficiency of XML is that it's verbose and requires significant resources, resources that many embedded devices find difficult to provide.

15.2 MTCONNECT

Those of us in the world of discrete and process automation think that everyone is either in the discrete world of individual parts or in the process world of continuous processes. The beliefs are that everyone uses PLCs and DCSs to control their systems, and that their network architectures are composed of the technologies most familiar to us: Modbus, PROFIBUS, DeviceNet, EtherNet/IP, and PROFINET IO.

That is a very limited view. There is a whole other world of automation devices—a pretty big one—that uses very sophisticated automation systems to generate products. That "world" is the machining or automated tooling domain, a world with sophisticated cutting and machining tools, specialized processes, and its own terminology and methodologies.

MTConnect, which started in 2006, is the communication standard in the machine tool industry. Prior to MTConnect, shop floor machine monitoring of a series of disparate tools from different vendors was impractical for all but the largest, most profitable, and most technologically adept machine shops.

Overview

In contrast to something like OPC UA, MTConnect is a pretty simple technology, one built on two very well-known standards: XML and HTTP. Very simply, MTConnect can be thought of as a well-defined standard for sending machine shop floor data as XML files. MTConnect uses simple HTTP Get instructions, the same instructions used for web servers to deliver web pages, to request machining data from a controller.

This works very well in the tooling industry, as the industry is small enough to define a common vocabulary and the information model semantics to serve most, if not all, of the information transfer requirements in the industry. Terms like power on/off, spindle speed, axis position, axis speed, feed, block number, status, CNC mode, work number, alarm, spindle load, axis load, and spindle override are all common to most of the equipment in the industry.

MTConnect separates the software functionality needed to transfer data from a machine controller to an application into two components: adapters and agents. Adapters are software components that interact with the machine controller. An adapter translates the machine controller data from its values and format (often proprietary) into the common terminology of MTConnect.

Agents are software devices that interrogate adapters. Agents have two functions: (1) collect data from one or more adapters and (2) provide that data to applications. Agents communicate with adapters in any communication mechanism available from that particular adapter. The MTConnect standard does not define how data moves from adapter to agent.

Agents are the software components that respond to HTTP GET requests from applications with XML files. Those requests can include PROBE requests, which are requests for the XML Schema for a machine controller, or STREAM requests, which are data requests. The XPATH language is supported, which provides the capability to limit requests to specific values.

Summary

MTConnect is an open, royalty-free communications standard intended to foster greater interoperability between machining equipment and software applications on the shop floor. Those goals are similar but less encompassing than the goals of a standard like OPC UA.

MTConnect is the standard for the machining world. IoT applications in machining would most likely be based on MTConnect directly or use a gateway that can receive MTConnect data and bridge it to another more enterprise-friendly IoT protocol.

15.3 HTTP

The Hypertext Transfer Protocol (HTTP) is the connectionless, stateless protocol[1] that is used every time we access a web page. It is included here because more than a few vendors implement it as a very simple way of moving data between automation devices and IT and IoT applications.

1. A stateless protocol is one that carries no information from previous transactions. Every message is its own completely standalone transaction. Protocols that carry state information respond differently to a message based on some state left over from a previous transaction.

What Is HTTP?

HTTP is a request-response protocol. A client establishes a TCP connection with a server and sends an HTTP request to the server. The request generally includes a URL, the protocol version, and a message containing request parameters, client information, and sometimes a message body. The server responds with a status line that includes the protocol version, a response code, and a message body.

Unlike many other computer protocols, HTTP closes the connection when the request is complete. A new request means that another connection must be opened. This requirement to open a connection on each request and not carry over any information from a previous request is what makes HTTP a stateless protocol.

An HTTP message contains either a GET request to retrieve information from the remote system, a PUT (or POST) request to send information, or a HEAD request, which returns everything the GET request does except the message body.

In its most often used implementation, the HTTP GET service is used to request a web page from a remote server. The response message contains the Hypertext Markup Language (HTML) that forms the web page content that appears in your browser.

How Is HTTP Used?

HTTP is a very simple technology that many vendors use to build applications that move data from automation devices to IT applications. It is not difficult for vendors to customize HTTP GET and POST messages and to add custom protocol information in the message body. By building applications that use these protocols, these vendors create easy-to-use mechanisms for moving automation data between IT systems and the factory floor.

For example, there is a device vendor that makes generic networking modules. This vendor supplies proprietary hardware and software for building networked applications and users write C code to build applications that interface with the vendor's hardware. By adding an API for a cloud application that uses HTTP, the user can easily move data from the hardware to applications in the cloud that also use the vendor's API. Users who are not concerned about using proprietary technology can easily create cloud applications that manipulate and store data from the embedded hardware.

Summary

HTTP, unlike some other IoT protocols, provides no information model, no services other than the raw GET and PUT, and no standardized mechanism to move data. Most implementations based on HTTP use a proprietary body that encodes commands and responses. Despite these limitations, HTTP's popularity and simplicity make it a very popular protocol for IoT applications where building a proprietary implementation is undesirable.

15.4 REST

Unlike the other concepts described in this chapter, Representational State Transfer (REST) is not a protocol or a technology; it is actually an architectural concept for moving data around the Internet. The REST architecture or a RESTful interface is simply a very flexible design, usually using the services of HTTP, for client devices to make requests of server devices using well-defined and simple processes.

What makes REST so significant is its widespread acceptance for many important applications as a simpler alternative to Simple Object Access Protocol (SOAP) and Web Services Description Language (WSDL). Leading companies like Yahoo, Google, and Facebook have passed beyond SOAP and WSDL-based interfaces in favor of this easier-to-use, resource-oriented model to expose the web services they offer to other applications on the web.

Not many people are aware that every time they browse a web page they are using the REST architecture.

What Is REST?

In REST, the concept of how devices on a network function is different than the conceptual view of a network for most other networking technologies. We usually think of a network as a set of devices that provide a specific set of functional services. A Modbus device, for example, provides a specific set of services like Read Coil, Read Holding Register, Write Coil, and Write Register Single. EtherNet/IP adapter devices and other Common Industrial Protocol (CIP) devices provide services like Read Attribute and Write Attribute. In most technologies used in industrial automation, there is a set of predefined functions (or services) that client devices must learn, implement, and use to access the resources of a device.

This function-specific architecture works well in tightly controlled systems that serve smaller problem domains. For many years, Modbus was the accepted way to pull data out of energy meters, using the Read Register command to get values from energy meters. In well-defined paradigms like energy data collection, these kinds of architectures make sense.

In these architectures it was just accepted that if a server device (like an energy meter) implemented a new service (like the capability to track detailed energy usage over a small period of time in great detail), all the clients that wanted access to that new resource would be adapted with new software to use that feature. That sort of architecture works well in these limited paradigm systems but it doesn't work well in the World Wide Web where vast amounts of unlimited, ever-changing resources are being made available.

It would be massively impractical to update our web browsers every time a change is made to a web page and the reason we don't have to do that is because the interaction between your web browser and the web server is a RESTful interface. Your web browser does an HTTP GET which retrieves markup text containing text and hyperlinks to other web pages. You now know more about the resources offered by that server and can request more information by clicking on one of the hyperlinks. That click repeats the process (another HTTP Get is issued, a new set of markup text is delivered) and you have access to more pertinent resources.

This works because a different mindset is in place about servers and the resources they offer. This mindset is resource-centric instead of function-centric. In the RESTful architecture, a server is viewed as a set of resources, nouns if you will, that can be operated on by a simple set of verbs like GET, POST, UPDATE, and the like. This architecture yields a much more flexible mechanism for retrieving resources than the limited function-centric kinds of technologies we've used in the past.

There are a few important principles that are key to understanding the RESTful architecture:

- REST is an architecture style—all web interactions can be said to be REST architecture operations.
- REST is not a standard—there is no specification for it from a standards body.
- REST is a stateless client-server protocol. Stateless meaning that the server remembers nothing from any previous interaction with that client.
- REST is based on HTTP but it can be implemented on other protocols that provide connection services.
- RESTful applications use the simple HTTP services called CRUD (Create/Read/Update/Delete).
- Like other web services, REST has no built-in security, it offers no encryption, and it does not do session management or any other added-value service. These can be made available by adding components to the transport or by using a different transport like HTTPS.

How Is REST Used?

Yahoo and Facebook have created RESTful client APIs which simplify the process of accessing resources within their systems. These APIs have largely replaced the remote procedure call technologies previously used, like SOAP and WSDL. In fact, it's clear that all these types of remote procedure call technologies are obsolescing and taking their place in Internet history.

But, you might say, it is clear that the REST architecture is perfect for browsers and humans accessing web pages, but how is it used for machine-to-machine (M2M) kinds of communication? That is one of the reasons that HTTP is so perfect for a REST architecture. REST has a built-in feature that allows the client to select the format in which the server should return the resource. Web browser clients use the resource to return markup language (HTML) that can be displayed. Machines, on the other hand, can request things like Java objects, XML files, CSV files, and other data more easily processed by machines.

Summary

REST is a very good alternative for building simple IoT applications. It is simple to understand and easy to implement, but less functional than some of the other alternatives. As a simple mechanism to move factory floor data to an IT application or cloud server, REST can be a good choice. You can implement a factory floor server that provides a REST interface, and define Java objects, XML, or CSV as the delivery format for your data. It won't be real time, but you don't always need real-time data.

15.5 MQTT

Overview

Message Queuing Telemetry Transport (MQTT) is another mechanism for moving data around the factory floor or from the manufacturing environment to the cloud. MQTT is designed to meet the challenge of publishing small pieces of data in volume to lots of consumers' devices constrained by low-bandwidth, high-latency, or unreliable networks. MQTT supports dynamic communication environments where large volumes of data and events need to be made available to tens of thousands of servers and other consumers.

What Is MQTT?

The heart and soul of MQTT is its publish/subscribe architecture. This architecture allows a message to be published once and go to multiple consumers, with complete message decoupling between the producer of the data/events and the consumer(s) of the messages and events.

Data is organized by topic in a hierarchy with as many levels of subtopics as needed. Consumers can subscribe to a topic or a subtopic. They can also use wildcards to specify topics of interest.

A broker receives the information from the servers and matches the information to consumers by topic subscriptions. The information is distributed to every consumer that matches the topic. If no consumer requires the information, the information is discarded.

Topics are designated by a namespace. Subtopics are designated with a slash ("/"). For example, an energy system might publish information about itself on:

<HouseID>/<system>/<meternum>/energyConsumption

where

HouseID	=	a specific location
system	=	the HVAC, Kitchen, or solar energy system
meternum	=	a specific meter in a system

Consumers can subscribe to all meters in a system with messages such as <HouseID>/Kitchen/* or all systems with <HouseID>/*.

What Are the Benefits of MQTT?

Some of the benefits of MQTT technology include:

- It efficiently generates event-based messages. MQTT is a "PUSH" system in which the producers push data to brokers. No bandwidth is consumed by consumers requesting data.

- Real-time or pseudo-real-time data is immediately pushed to consumers with a minimum of delay.
- The resource requirements for publishers are reasonable. MQTT is a very good choice for low-resource devices like sensors and actuators.
- It facilitates reliable operation on fragile and unreliable networks. Brokers can be configured to retain messages for consumers that are temporarily disconnected.

Summary

MQTT is a very simple way of distributing information from lots of publishers to lots of consumers. The ability to support thousands of devices distinguish it from some of the alternatives already discussed. It is extremely lightweight, reliable, and adapts well to low-resource devices. Broker devices, which some view as a disadvantage, manage the connection between the publishers and consumers.

MQTT is a superior way of moving data from sensors and actuators to various kinds of consumers. The publish/subscribe technology of MQTT is vastly superior to that of other technologies, as MQTT is designed to minimize network bandwidth especially in applications that have thousands of publishers simultaneously reporting data and events.

15.6 DDS

Overview

The Data Distribution Service™ (DDS™) is an architecture for data-centric connectivity.[1] DDS provides more complete communication services than other IIoT (Industrial Internet of Things) communication systems including automatic data discovery, security, content scheduling, and other management tasks that are normally provided to applications in other technologies.

DDS is designed to manage a large network of publishers and subscribers by focusing on the data being communicated instead of the communication method. Instead of messaging, DDS focuses on the contextual relationships of the data, and organizes publishers and subscribers around the data they need and when they need it.

Unlike many other IIoT protocols, DDS completely decouples subscribers from publishers. Geographic location, redundancy concerns, time sensitivity, data distribution, and platforms are all decoupled by the DDS architecture.

What Is DDS?

At the core of DDS is the concept of data centricity. Data appears to each application simply as a local store. An application interacts with that local data store using its native data model. A Java application, for example, works with Java objects. The application uses the data as if it is not part of the IIoT. DDS manages that local data store and connects it to other publishers and subscribers of data on the network without specific guidance from the application.

Data is organized by topic within a domain. Domains are a group of related topics. A data writer and a data reader can only match topics if they join the same domain. Topics can be shared or exclusively owned. There can be multiple topics and instances of topics. Topics can be filtered as needed so information flows only when it is relevant and necessary.

---

1. Data-centric connectivity refers to the connectivity of software architectures that are organized around data table, databases, or other kinds of data storage mechanisms. Data-centric architectures generally focus on the data, not on logic or messaging.

Like MQTT, OPC UA, and other technologies, DDS uses a publish/subscribe communications model. Data writer applications write data into their local data store that DDS makes available to data readers that subscribe to topics available in the domain.

DDS does this automatically by matching data writers and data readers. Consumers, or subscribers, interested in a topic can declare an interest in that topic. DDS matches that interest to a publisher or data writer offering that data. The matching of writers to readers is entirely managed by DDS and not the application code. It is an anonymous arrangement. The data writer has no information on how many data readers, if any, are interested in its data.

There are two architectural elements to DDS: Data-Centric Publish-Subscribe (DCPS) and Data Local Reconstruction Layer (DLRL). DCPS manages the exchange of topic data with other DDS-enabled systems. DLRL manages access to topic data from the local application.

What Are the Benefits of DDS?

Some of the benefits of the DDS technology are:

- DDS can handle timing-critical distributed systems better than any other IIoT competitor.
- Applications using DDS are decoupled from traditional network interactions like determining who should receive messages, when they should receive them, and what to do if something goes wrong.
- DDS automatically matches writer to reader without guidance from the application.

- Applications only interact with what they perceive as a local data store.
- Applications interact with the local data store in their native data format.
- DDS is very tolerant of network delays.
- DDS offers low latency. Information is immediately pushed to consumers.
- Content can be filtered by the DDS network to eliminate data that is not needed by a subscriber.
- DDS delivers very reliable operation on fragile and unreliable networks.

Summary

DDS is a device-to-device architecture. It is an IoT protocol that simultaneously moves millions of messages per second to numerous receivers. Though DDS is capable of moving data to IT applications, it isn't designed for that.

DDS is very impressive technology. Its publish-subscribe architecture appears to be much better than that offered by OPC UA because of its device-to-device design, which enables it to easily integrate a large number of nodes into a domain and share topic data.

DDS with its efficient bandwidth management, simple application interface, and low overhead is one of the best technologies for delivering time-sensitive data and high-performance device-to-device messaging.

Index

1000BASE-CX
 Copper-Twinax Cabling 36
1000BASE-LX
 Vertical or Campus Backbones 36
1000BASE-SX
 Horizontal Fiber 36
100BASE-T 30
10BASE2
 Thin Ethernet (Thinnet) 26
10BASE5
 Thick Ethernet (Thicknet) 26
10BASE-F
 Fiber-Optic Ethernet 29
10BASE-T
 Twisted-Pair Ethernet 27
2-Port Routers 92
7-Layer Networking Model 45

Access Points 160
Access Routers 93
Addresses 70
Aloha Net 57
Amazon 177
AMQP 176
Application Services 179
Arbitration Mechanisms 21
ARP 83
ARP Utility 84
Attachment Unit Interface (AUI) 38
Attenuation 9
Auto-Negotiation 39

Bandwidth 9
Baseband 13
Bridge 89
Bridging Router (Brouter) 93
Broadband 13

Cables 115
 Installation 116
Cables and Connectors 116
Checksum 14
Chemicals and Temperature 116
Classless Inter-Domain Routing (CIDR) 70
Connectors 49
Contention 21
CSMA/CD Protocol 42

Cyclic Redundancy Check (CRC) 14, 60

Daisy Chain Topology 21
Data Distribution Service™ (DDS™) 210
DDS 210–213
Deployment 179
Determinism and Repeatability 94
Devices 88
DHCP 76
Digital Communication Terminology 9
DNS 78
Documenting 105
Drivers and Performance 98
Duplex 15

Electromagnetic Interference (EMI) 113
Error Detection 14
Ethernet
  Fast 30
  Grounding Rules 111
    for Coaxial Cable 111
  Hardware
    Basics 25
    LEDs 37
  Terminology 25
Ethernet and TCP/IP
  History 4
Ethernet Building Blocks 87
Ethernet DB-9 Connector 54
Ethernet Design Rules 45
Extensible Markup Language 197

Fiber-Optic Connections 118
Fiber-Optic Distance Limits 118
Foundation Services 178
FTP 78
Full-Duplex 16
  Ethernet with Single-Mode Fiber 120

Gateways 94
Gigabit Ethernet 35
Gigabit Media Independent Interface (GMII) 38
Global Infrastructure 178
Grounding for Shielded Twisted Pair 113

Half-Duplex 16
HTTP 78, 202–204
HTTPS 176
Hub/Spoke or Star Topology 18
Hubs 88
Hypertext Markup Language (HTML) 203
Hypertext Transfer Protocol (HTTP) 202

IEEE 151
IEEE Standard 802.11 157
IEEE Standard 802.3 57
Industrial Ethernet 1
Installation, Troubleshooting, and Maintenance 111
Intelligent Bridges 90
Interface Cards 94
Intermediate Interface 38
IoT 175–177, 179, 197
IoT Hub 176

LAN vs. WAN vs. VPN 22
Layer 1, Physical Layer 48
Layer 2, Data Link 48
Layer 3, Network 48
Layer 4, Transport 47
Layer 5, Session 47
Layer 6, Presentation 47
Layer 7, Application 46
Legacy Address Classes 67
Loosely-Coupled Systems 185–186

M12 "Micro" Connector 55
Magnetics 38
Managed Hubs 89
Management 179
Media Access Controller (MAC) 37, 60
Media-Independent Interface (MII) 38
Mesh Networks 161
Mesh Topology 20

Message Queuing Telemetry Transport (MQTT) 208
Metcalfe, Robert 57
Monitoring 102
MQTT 176, 208–210
MTCONNECT 200
MTConnect 201–202
Multi-Port Routers 92

Netstat 83
Network
  Collisions and Arbitration 39
  Health, Monitoring, & System Maintenance 101
  ID vs. Host ID 67
  Time Servers 94
Noise 10

Objects 188
OPC Classic 189
OPC UA 188–189
OPC Unified Architecture (UA) 188
Oracle 179

PC-Based Ethernet Utilities, Software, and Tools 108
PHY 38
Physical/Embedded Components
  MAC, PHY, and Magnetics 37
PING 80
Pinouts 52
Polling 22
Power over Ethernet (PoE) 151
Power Sourcing Equipment 152
Powered Device 152
Priority Messaging 97
Protocol vs. Network 15

Repeaters 89
Representational State Transfer (REST) 204
REST 204–208
RESTful Interface 180
RFID Systems 151
Ring Topology 19
Routers 92
  Types of 92

Security 161, 178
Segmented Hubs 89
Signal Encoding Mechanisms 11
Signal Transmission 9
Signaling Types 13
Simplex 15
SNMP 77
Stacking or "Crossover" Cables 89
Switches 91, 97
Switches vs. Hubs 115

TCP/IP
  Application Layer Protocols 76
  Utilities 80
TCP/IP Protocol Suite 61
  IP Protocol 61
  TCP Protocol 61, 71
  UDP Protocol 74
Telnet 79
Terminal Servers 93
TFTP 77
Thin Servers 93
Tightly-Coupled Systems 185
Token 22
Topologies 17
Transition 12
Transmission/Reception of Messages 15
Troubleshooting 106, 111
Trunk/Drop (Bus) Topology 20
Twisted-Pair-Cable Types 112
Two-Speed Hubs 89

Wireless Ethernet 157
Workgroup Hubs 88

XML 197–200